The Day
LINCOLN
WAS
SHOT

The Day
LINCOLN
WAS
SHOT

★ ★ ★

JIM BISHOP

HARPER ● PERENNIAL

NEW YORK ● LONDON ● TORONTO ● SYDNEY ● NEW DELHI ● AUCKLAND

HARPER PERENNIAL

A hardcover edition of this book was published by Harper & Brothers in 1955.

THE DAY LINCOLN WAS SHOT. Copyright © 1955 by Jim Bishop. All rights reserved. Printed in the United States of America. No part of this book may be used or reproduced in any manner whatsoever without written permission except in the case of brief quotations embodied in critical articles and reviews. For information, address HarperCollins Publishers, 10 East 53rd Street, New York, NY 10022.

HarperCollins books may be purchased for educational, business, or sales promotional use. For information, please e-mail the Special Markets Department at SPsales@harpercollins.com.

First Perennial Library edition published 1964 by Harper & Row, Publishers, Incorporated.

First Harper Perennial edition published 2013.

Library of Congress Cataloging-in-Publication Data is available upon request.

ISBN 978-0-06-229060-1 (reissue)

13 14 15 16 17   OV/RRD   10 9 8 7 6 5 4 3 2 1

Dedicated to My Dearest Friend
John M. Bishop
Who Is Also My Father

Contents

CONTENTS

Good Friday was the day
Of the prodigy and crime,
When they killed him in his pity,
When they killed him in his prime. . . .
. . . They killed him in his kindness,
In their madness in their blindness,
And they killed him from behind. . . .

He lieth in his blood—
The Father in his face;
They have killed him, the forgiver—
The Avenger takes his place. . . .

There is sobbing of the strong,
And a pall upon the land;
But the People in their weeping
Bare the iron hand:
Beware the People weeping
When they bare the iron hand.

—*Herman Melville*

For the Record

This is a book about a day, a place and a murder—and about a wide variety of men and women. It begins with the casual and somewhat late good morning of President Abraham Lincoln outside his bedroom door at 7 A.M. on Friday, April 14, 1865, and it ends at 7:22 A.M. the following morning when, as Surgeon General Barnes pressed silver coins to the President's eyelids, Mrs. Lincoln moaned: "Oh, why did you not tell me he was dying!"

The elapsed time is twenty-four hours, twenty-two minutes. To many people, this is the single most dramatic day in the life of the Republic. It has been written about before, as a chapter of a book, or a part of a chapter and once in a terse volume written by John W. Starr—a volume which ended at 8 P.M. Some of these passages have been beautiful and moving and some have been skimpy and vague and laden with unsupported suspicion. Some of the lurid journalists, feeling that there was not sufficient natural drama in the violent death of Lincoln, filled in the blank spots of this day with imaginings, and the story of the assassination, in time, became so interlarded with fiction that the principal assassin, John Wilkes Booth, became a minor character.

In addition to the chapters dealing with specific hours of the day and night, I have included here two chapters of background in a section entitled *The Days Before*. I have been reluctant to interrupt the narrative with the insertion of this background section, but have been persuaded that this is necessary and useful in placing the events of the day in context.

As a student of President Lincoln and his times, I began, in 1930, to keep notes on the events of this day. The best, and simplest way, I felt, was to keep notebooks labeled 7 A.M. Friday, 8 A.M. Friday, 9 A.M. Friday and so on through 7 A.M. Saturday. That made twenty-five notebooks. In addition, I kept one marked "Lincoln and Family," one labeled "The Conspirators," one called "Washington—Era," and one marked "Bibliography." This must be of small interest to any reader except to point out that, after years of reading and making notes, I found that I had as many as three or four versions—each at variance with the others—of what had happened in any one hour. Two years ago, when I intensified the research and started to read seven million words of government documents, the pieces of the puzzle began to orient themselves. There were still conflicts of time and place and event, and these were eventually reconciled by (1) the preponderance of evidence tending toward one version: (2) the testimony of more than one supporting witness at the trial of the conspirators: (3) the relationship of the event in question with an event that occurred prior to it or immediately after.

Still, I do not believe that this book presents all of the facts, nor anywhere near all of the facts. In the little notebooks today, hundreds of pages are marked "void." In the multitude of trial records, documents and books, there are many blank places although, to compensate for this, I must acknowledge that many of the witnesses supplied sufficient material so that conversations could be reconstructed in dialogue without straining the quotation marks. In fact, the only liberties I have taken are in describing facial expressions ("he scowled"; "Booth looked tired," etc.) and in describing what certain characters thought, although in each case the thought is based on knowledge of facts then in the possession of the character. Other than that, this book is pretty much a journalistic job.

To insure that the book should be factually sound, galleys were sent to such Lincoln scholars as Bruce Catton, Stefan Lorant, and Harry E. Pratt, Historian of the Illinois State Historical Library. I hope that their suggestions and their challenges have been met and that the book is better, smoother and more sure of itself because of their ministrations.

Sometimes, small facts become elusive. For example, I assumed all along that Lincoln's office was on the ground floor of the White House. It did not occur to me to challenge this until I read in a Carl Sandburg book that the President was informed, in his office, that people "were waiting downstairs." The book was nearly complete when, through the kind offices of Congressman Frank Osmers (N.J.), it was put beyond dispute that Lincoln worked upstairs, not down. Another "small" fact is that most writers assumed that Booth, in his escape from the alley behind Ford's Theatre, spurred his mare up the alley to F Street, and turned right. It did not occur to me to question this until I learned, in an old document, that a wooden gate, used as a billboard, closed the F Street exit and that the assassin would have had to ride up the alley, halt, dismount, open the gate, and then flee. In Ford's Theatre, a National Parks guard told me that the alley, in 1865, formed a T, and that John Wilkes Booth was aware of the gate at F Street and had not used it, turning instead down the other leg of the alley to Ninth Street, and thence right to Pennsylvania Avenue. In the library at the back of Ford's Theatre, this guard had an old government pamphlet which proved the point.

For help in amassing the material for this book, I am indebted to Mr. Evan Thomas, Managing Editor of Harper & Brothers; Mrs. Phyllis Jackson of Music Corporation of America; Miss Olive Tambourelle of the Teaneck, New Jersey, library; Mr. Robert Hug of the New York Public Library; the Illinois State Historical Society; to the Esso people for a fine street map of Washington; to Gayle Peggy Bishop, age ten, for facing thousands of pieces of carbon paper in one direction; to Virginia Lee Bishop, seventeen, for believing that no one but her father could have written this particular book.

Jim Bishop
Teaneck, New Jersey

Daybreak

★ ★ ★

7 a.m.

The polished rosewood door swung back and the President of the United States came from his bedroom. He nodded to the nightman in the hall and said "Good morning." He fingered his big gold watch, anchored to the chain across his vest, but he did not look at it. The hour of seven was late for Lincoln. Many a time, the guard remembered, the President was downstairs working at six.

The big man started down the hall slowly, like a person older in years, the legs perpetually bent at the knees, the black suit flapping about the frame. He looked like a man who did not feel well. The circles under his tired eyes were pouched; the skin of his face was almost saffron; the scraggly black beard thinned and died as it approached the hairline; the hair itself was almost combed; the feet moved with conscious effort, barely lifting off the red pile rug before being set down again; the thick lips, more brown than red, were pulled back in a semi-smile.

He saw the men ahead. There was no way to avoid them. The guards could not seem to keep them out, and many of them slept in the White House hall. The word had passed that he was coming, and so they were on their feet and smiling. Each of these wanted a favor. As he passed, hardly pausing, they asked for jobs or passes to Richmond or the commutation of a military sentence or presidential approval of an illegal business deal. In four years of living in the White House, Mr. Lincoln had become accustomed to the morning vultures. He could do little to be rid of them, and he had no desire to help them because, if their claims were just, they would have had satisfaction at the proper agency.

There was no way around them. His bedroom was in the southwest corner of the White House, on the second floor, and his office was in the southeast corner, on the same floor. Some men, desperate or arrogant, grabbed the crook of his arm and held him until the President pulled himself loose and said: "I am sorry. I cannot be of help to you." Some spoke quietly and swiftly, their heads swinging to follow him as he kept walking. Some wept. A few muttered threats and departed.

He walked down the carpeted hall slowly, and through the door to his office. A soldier came to attention, and Lincoln nodded pleasantly and walked inside. It was a big office, bigger than two farm kitchens, and he walked over to the far side from the door and looked in the pigeonholes of the old desk and then sat at the small table near the south windows. He picked up a paper, crossed his legs and leaned back in the high chair. The light from the windows was not good; this was a misty morning and a chill breeze leaned against the newborn buds in Washington City. This would be a day for a coat in the morning hours and the evening hours. Down behind the Capitol, with its new dome looking like a marble breast, the sun was fighting a patient battle with gray clouds.

The streets, at this hour, were full of people. Unofficial Washington was on its way to work. In offices and shops, the business day began at 7:30 and the flagstone and wooden walks rang with the tempo of heavy boots. In the gray clay, teams of horses steamed as they pulled heavy brewery wagons and loads of produce to the taverns and markets of a gluttonous town. The drivers, in open vests and black peaked caps, bounced as they drove through brown puddles and kept an eye on the meticulous ladies who dipped brooms in hot water to sweep mud from the steps of the houses.

Washington City was a place of cobblestones and iron

wheels, of hoop skirts and gaslight, of bayonets and bonnets, of livery stables and taverns. It was a city of high stoops and two-story brick houses with attics. From almost any point in the city, the dominating features were the Capitol and the Washington Monument. With a thumb on the dome of the Capitol, and a middle finger on the Monument, a circular span would embrace practically all of Washington City.

The town was Southern in character and habit, leisurely enough so that individual noises could still be heard. On Sunday mornings, no matter how deeply one huddled in blankets, bells from at least two churches could be heard tolling. A train leaving the Baltimore and Ohio terminal would shriek once, and a hundred shrieks would come back from the streets. A lounger at Pennsylvania Avenue and Third Street could watch a truck pass by, meet a friend, have a conversation, stop for a drink, and still return to the corner in time to see the same truck far up Pennsylvania Avenue, perhaps at Thirteenth Street.

It was a busy town, but compared to New York or Philadelphia or Chicago, or even compared to its neighbor to the north, Baltimore, it was small and pompous. Below Gravelly Point, ships under sail were standing in to port, passing paddle-wheel steamers, new in white paint and gold-leaf trimming, squatting low in their own lace train.

It was a Currier and Ives print come to life, and more. It was a city of individual persons with unique passions and ambitions. The popular song was "When This Cruel War Is Over." Women bought the freshest editions of the newspapers to study the columns of war dead, which started on the left side of page 1, and then jumped to an inside page. Each day they performed the simple, breathless duty of looking. With a finger on type, they moved from column to column until they found a heading marked "Ord's Corps" or "Sheridan's Army," and then Mrs. Jones moved her finger down to the J's, expelled a big breath, and began to read the news.

The people, by later standards, were adolescent and did much of their thinking with their hearts. They were emotional and gullible, and morbidly concerned with the imminence of death. The latest intelligence from abroad, which came aboard a packet boat just landed at Castle Garden in New York, was that the Duke of Northumberland had died and Cardinal Wiseman was not expected to live. Anyone who had consumption could hardly do better than to buy Dr. Wishart's Pine Tree Tar Cordial, and Dr. Morris advertised "a secret worth knowing to married females." Another reputable doctor offered to cure "Cancer for $2 a visit—no cure no pay." A tooth could now be extracted "without pain, with nitrous gas, ether, or chloroform" for 50 cents. The Baltimore Lock Hospital advertised itself as a "refuge from quackery; the only place where a cure can be obtained." Among the ills corrected were "weakness in back or limb, involving discharges, impotency, confusion of ideas, trembling, timidity—those terrible disorders arising from the solitary habits of youth." The hospital's boast was: "The doctor's diploma hangs in his office."

Most grocery stores were really general stores. They sold groceries, meats, wines, liquor and hardware. Fresh geese and hares hung from barrels in the doorway. Prices were a wartime outrage. Firkin butter was 30 cents a pound, coffee was hard to get at 21 cents, salt cost 50 cents a bushel, corsets cost $1.50 (extra strong ones $1.75). Hoop skirts wholesaled for $1 apiece and figured prints retailed for 15 cents a yard.

The anguish of housewives was met by the complacent shrugs of the merchants, who denied that outrageous profiteering was ruining the American dollar. They now had plenty of merchandise, the bins, barrels and jars were filled to brimming with flour, crushed meal, brown sugar, green and black tea, spices, sauces, jellies, starch and yeast, tobacco, cigars and snuff, oil of coal, sperm and ethereal, kerosene lamps, marble tabletops and foot warmers. The favorite whiskeys were Baker

1851, Overholtz 1855, Ziegler 1855 and Finale 1853. Holland gin was sold loose from barrels. The highest-priced meats were ham, at 28 cents a pound, and turkey at 30. A barrel of Boston crackers, enough to last a season, cost $6.50.

Bricklayers were getting $2.50 for a day's work, and demanding $3.50. Freed slaves were paid $11 a month and keep for field work. A stranger walking through Washington City would believe, from what he saw, that the main businesses of the town were livery stables and wood yards. There seemed to be one of each to each block, plus a tavern on the corner. The smell of wood smoke was the cologne of the streets. Ducks and chickens picked along Pennsylvania Avenue, edging without panic around the horses, and pigs wallowed and grunted in the street puddles.

The White House was big and shabby. Successive Congresses had refused to repair it. The rugs were patchy and thin from traffic and mud. The drapes were ornate, but souvenir hunters had cut swatches from them and had stolen silverware and even snuff boxes. In good weather, the odor from the canal on the south was sickening and the malarial mosquitoes were belligerent.

The building faced Pennsylvania Avenue, its white columns glinting gold in an afternoon's sun. An iron paling fence kept the curious out, but paths to various government departments transversed the lawns. The south grounds, facing the Potomac River, had stables, outhouses and work buildings. Squatters tented on these grounds, and little Tad Lincoln kept goats. The building was flanked by the State Department, the Treasury Department, the War Department and the Navy Department, all on ground marked "The President's Park."

South and east of the White House and the Capitol were acres of Negro shanties. These people, newly freed, had come north to be treated as people; there were too many of them and the labor market was so depressed that Negro women often went from house to house offering to work for anything, for food for their children.

North of Rhode Island Avenue were stands of timber and farms. Georgetown, except for its university, was largely farm land and suburban dwellings. The city—the real city—lay between Capitol and White House and across a few streets to the north. Pennsylvania Avenue was called "the Avenue" and it had one sidewalk, on the west side. On the other side were open markets and a drainage ditch. The main buildings, excepting the White House and the Capitol, were the Post Office, the Patent Office, the Treasury and the Smithsonian Institution.

Horse cars swayed slowly down the Avenue, and connected with the Sixth Street cars at East Capitol for the Navy Yard. Convalescent soldiers slept in the basement of the Capitol, vying for space with nearly ten thousand tons of flour which had been cached in case of siege. The railroad terminal, an ornate wooden station with tracks on the street level, stood three blocks north of the Capitol and the pride of the Baltimore and Ohio was that it could whisk a statesman to Baltimore (forty miles away) in an hour and three quarters, or get him to New York City in nine hours.

Hotels were an innovation to the city, which had been accustomed to boardinghouses and taverns as homes-away-from-home. The lobbies and sitting rooms were alive with traffic. Ornate chandeliers hissed with gaslight, and the door-men and servants, in uniforms black and maroon, eased the lives of legislators and their wives. Official Washington liked the hotels at once. The bars were crowded, the carpeting rich, the toilets were indoors and the spittoons glittered.

The Willard, at Fourteenth and E, was considered by the fashionable set as *the* place to be seen. The National, at Sixth and Pennsylvania Avenue, catered to Southerners although not exclusively so because, on this morning of April 14, the hotel registry showed that among the guests were ex-Senator John P. Hale of New Hampshire, and his family; John Wilkes Booth, an actor, of Bel Air, Maryland; and Speaker of the House Schuyler Colfax, of Indiana, who favored a harsh peace for the South. Negro servants leaned against these buildings in giggling chatter while their masters, in long fawn-colored coats and umbrella-brimmed hats, transacted their business inside over a mahogany bar.

Brown's, across the street from the National, was another good hotel. So was the Kirkwood, where Vice President Johnson stayed, and Herndon House.

Among the more permanent institutions were a penitentiary, twenty-four military hospitals, an insane asylum, a huge poorhouse, an assortment of low- and medium-priced houses of prostitution and a score of publicly acknowledged gambling houses.

The Washington police force consisted of fifty policemen who worked by day and were paid by Washington City, and a night force of fifty more who were paid by the Federal Government. The night men were not paid to protect citizens; their job was to protect public buildings. The Fire Department was paid by the city, but it was controlled by politicians and often refused to go out to fight fires. The criminal code of the District of Columbia was archaic and was enforced largely on political grounds. Crimes punishable by death were murder, treason, burglary, and rape if committed by Negroes. Only a few years before this day, many of the politicians who fought for the abolition of slavery made extra money by selling freedmen back into slavery. Until the Emancipation Proclamation had been signed in 1863, a weekly auction of Negroes was held in the backyard of the Decatur House, a block from the White House.

There was a great difference between "permanent" Washington and political Washington. A clerk earning $1,500 a year in the new Treasury Building found it difficult to feed a wife and children and his quarters were little better than what the Negroes had. He was at his desk at 7:30 A.M. and, in the evening, he left it after 4. Political Washington functioned between November and June, when Congress was in session. It convened late and it did not convene every day.

The hotels, which understood the legislators, served breakfast between 8 A.M. and 11. A good breakfast consisted of steak, oysters, ham and eggs, hominy grits, and whiskey. Dinner was served at noon and ran to six or eight courses. Supper was disposed of between 4 P.M. and 5. Teas were common at 7:30 P.M. and cold supper was eaten between 9 and 10 P.M.

It was a city of handsome women too, and stout women were most admired. Congressmen's wives had more license in their behavior here than at home. They spent more for bonnets and gloves and they were equipped with cartes de visite and dropped them on trays in all manner of homes. They thronged the galleries of both Houses of Congress and, if a husband was busy, it was considered correct for the lady to choose an escort for the day. Even middle-aged women engaged in flirtations, or matters more serious than flirtations, and sometimes these ended tragically.

Dressed, the ladies looked like great Christmas bells, and their carriages, surreys, gigs and coaches were seen everywhere. They seemed always to be en route to or from a social call. From the moment that the season opened, on New Year's Day, with eggnog and hot punch and a presidential handshake at the White House, until Congress adjourned in the late spring, every family had an at-home night per week and spent all the other evenings visiting, or attending the opera or the plays. Under cut-glass chandeliers, they danced and drank and ate late suppers.

Their special pet was William, who made bonnets in an exclusive shop on Pennsylvania Avenue. He understood the exquisite agony of a lady who must have a narrow velvet ribbon of puce for a certain bonnet, and who desired that the remainder of the roll of that ribbon be destroyed.

This day was Good Friday, the day on which Our Lord and Saviour Jesus Christ died. In religion and in history, it was a solemn day and, from dawn onward, the churches were peopled. In the Catholic churches, such as St. Patrick's and St. Aloysius's, the statuary and the Stations of the Cross were hung in purple. It was the last full day of Lent, and the first day on which the Civil War would be referred to in the past tense. It was over, done with, finished, and Washington had been drunk for a week.

At bars, in clubs, at home, men fought and refought the whole war, and won it every time. Shiloh and Antietam and Gettysburg and Cold Harbor would now become pages in history books. Spotsylvania and Vicksburg and Chickamauga and Bull Run would become sites for monuments and markers and picnics. In some of these places the dead were still grinning, and, in others, the broken arms of bridges still held an awkward pose. For a long time, the walls of hospitals would hear the night cries of men in pain and, among women, black would be a fashionable color.

Over 600,000 men North and South were dead under hyacinth and weeds and swale grass and rock. Their congealed blood glued the shattered Union, and 29,000,000 persons were alive to enjoy the fruits of brotherhood. The national debt was high, $2,366,000,000, but the national economy was firm. Money wasn't scarce. On this very day, a man could get $1,000 bounty for enlisting for one year in Hancock's Corps. Many thousands of draftees paid from $300 to $450 to buy the services of a substitute soldier. In the matter of slavery, 384,884 persons had owned 3,953,742 other persons. Only one man in all the land had owned as many as a thousand slaves.

The death lists still came in daily, although the war was over, except for Confederate General Joseph Johnston's exhausted army and a few smaller units farther west.

Some people would die at home on this day. These were the unknowns, the unremembered. Louis Druscher, after a long and tiring fight, expired in his thirty-fourth year. Kate Anderson, aged twenty-five years and twenty-nine days, would die after a lingering illness which she "bore with Christian fortitude." Olive Louise Brinkerhoff, ten months of age, strangled of diphtheria. Pretty Violetta, daughter of Major Thomas Landsdale of the Maryland Line, died suddenly.

Small stones dropped into small pools.

It was 7:30 A.M. and official Washington, and lazy, unofficial Washington began to come alive. The President still sat at the small table in his office, reading official correspondence, one leg across the other, the free foot flexing slowly in the air.

A few streets to the north, on K St. opposite Franklin Square, Edwin McMasters Stanton spooned his soft-boiled eggs and asked Mrs. Stanton to please send regrets to Mrs. Lincoln. He was not a theatergoer and he was not going to be a party to a spectacle at Ford's Theatre tonight. Countless times he had advised the President to stay out of theaters and to cut all public appearances to a minimum, but, in social matters, he had found that a Secretary of War carries less weight than a First Lady. He asked Mrs. Stanton to get the handyman to fix the front doorbell. It was of the pull type, and it was broken. He was in a hurry; he wanted to visit poor Seward before reporting to the War Department. He would need the carriage.

Mr. Stanton all his life wanted to be a strong, efficient man, and he was. His strength lay in his will and his tongue. He was a short, paunchy person who affected square, gold-rimmed spectacles and gray, scented whiskers and the impatient, fluttery attitude of a man who is always trying to catch a mental train. He made and broke men mercilessly and he often appeared to bend Mr. Lincoln to his will.

On at least one occasion, the two men locked horns and, when it was over, the President prevailed. Mr. Lincoln had issued an order, without consulting his Secretary of War, that all Confederate prisoners who wanted to fight for the Union were to be freed. Stanton seethed. He hurried to the White House and he confronted the President and, barely able to restrain himself, he pointed out that if the order was executed, the one-time Confederate soldiers would be wearing uniforms of blue and, if captured, they faced hanging.

"Now, Mr. President," he said angrily, "those are the facts and you must see that your order cannot be executed."

"Mr. Secretary," the President drawled, "I reckon you will have to execute that order."

"Mr. President," said Stanton, "I cannot do it."

Lincoln locked his teeth and said softly: "Mr. Secretary, it will *have* to be done."

It was.

The Secretary of War was a complex man who, in his spare time, was composing a book entitled: *The Poetry of the Bible*. He was loyal, stubborn, hardworking, pedantic, driving, emotional, hated. When he was angry, he closed disagreements with "That will do, sir!" Sometimes, in his hands, power and cruelty were synonymous. And yet, when the power was in other hands and used against him, Mr. Stanton cringed for mercy. He was a perfectionist, and a master of deceit. He abhorred lying but he practiced it.

At the time that General McClellan became a public hero, the Secretary of War took note of it, then announced that McClellan expected to take Richmond soon, even though the general's telegrams of that day proved that McClellan was fearful of being overwhelmed by superior forces. When Mc-

Clellan failed to take Richmond, the public blamed the general and the press called him a disappointment.

In war, Stanton favored a suspension of the writ of habeas corpus and, in his tenure of office, sanctioned the arrest of more than a quarter of a million persons. Once, when a colonel was refused a request, he stared at the secretary and roared: "You can dismiss me from the service as soon as you like, but I am going to tell you what I think of you." Stanton heard the man out, then granted the request.

Lincoln summed up his own feelings about his Secretary of War when he said: "Stanton is the rock upon which are beating the waves of this conflict . . . I do not see how he survives—why he is not crushed and torn to pieces. Without him, I should be destroyed."

Over at the Metropolitan Hotel, a group of Baltimore celebrants watched Michael O'Laughlin get out of bed hungover and sad of eye. They had come down from Baltimore yesterday to celebrate the end of the war. His friends, dressed, laughed as Mike sat on the edge of his bed in long drawers and tried to orient himself. He was a small, delicate-looking man with long black hair and heavy imperial mustaches. They asked him if he could stand a drink. He looked up, smiled sadly, and said that he would have one after he got a shave.

Mr. O'Laughlin was an old friend of John Wilkes Booth. He was one of the original band of conspirators. Now, he lived at 57 North Exeter Street, Baltimore, and he worked for his brother in the produce and feed business.

Across the street from the White House, on the east side of Lafayette Square, the oldest man in the Cabinet was shaved in bed in his room on the third floor front. He was William H. Seward, near sixty-four, Secretary of State. He was white-haired, almost patrician, and he was in constant pain.

Nine days before, on April 5, his span of blacks had become frightened and had run away with his carriage, smashing it against the curb. He was taken home unconscious.

Charles Wood, of the Booker and Stewart barber shop, was doing the shaving, and he didn't like the job. The secretary was on the extreme edge of the bed, so that his broken right arm could hang free, and the double iron brace around his neck and jaw forced Mr. Wood to use only the most delicate of strokes as he kept the secretary smooth-skinned. Mr. Seward could talk in lipless grunts, and he listened to his son Frederick, Assistant Secretary of State, explain what he planned to say, in his father's name, at the Cabinet meeting this morning.

Stanton came in, asked how Seward felt, related the latest good news, watched the barber clean his straight razor, and departed for the War Department.

On the other side of town, John Wilkes Booth got out of bed and washed. He had had less than six hours of sleep, but he was restless. He was vain, courtly and meticulous. He scrubbed carefully, rubbed scented pomade in his hair and on his black mustache, put on a fresh suit and riding boots, checked the money in his wallet, and stepped out of his room at the National Hotel. This man had earned $20,000 in a year as an actor. As a Southern patriot, he earned nothing.

He was not tall, but he had the lean and bouncy quality of a man ready to spring. His hair was as black as washed coal and his eyes had a liquid quality of articulation which found a quick, sympathetic response in many women. He was his mother's favorite, although he was her ninth child, and she called him "Pet." Secretly, silently, sometimes poutingly, she worried about him. At twenty-six, he was a fine horseman, an expert fencer, a crack shot. In the world of the theater, he was only slightly less known than his father, Junius, one of America's great Shakespearean tragedians, who had died twelve years before, and his brother Edwin, perhaps America's greatest Shakespearean actor. Wilkes had bowed legs, but he hid them with custom-made wide pantaloons and extra-long coats.

As he left his room on Good Friday morning, his friend George Atzerodt was registering for a room he did not want at Kirkwood House, Twelfth Street and Pennsylvania Avenue. Mr. Atzerodt was assigned to Room 126, at the head of the stairs, on the left side of the corridor, almost directly above the room occupied by Vice President Andrew Johnson. Atzerodt took the key to his room and walked out. He had a consuming ambition to get as drunk as possible as quickly as possible. His assignment today was to kill the Vice President of the United States.

Other men, good and bad, were getting out of bed. Down in Fort Monroe, Virginia, Samuel Arnold had been out of bed an hour. Once, like his friend Michael O'Laughlin, he had been a Confederate soldier. Now he was a store clerk outside a Union fortress. He was young and had brown curly hair and dark eyes. Sam liked to make friends. One of his friends was Wilkes Booth. In Sam's pocket was a letter from Wilkes asking him to come to Washington at once. Arnold knew what that meant. The old plot to kidnap the President was being revived. Sam wouldn't leave the fort. Nor would he answer the letter.

The stone-jawed William Dennison was at breakfast. Formerly Governor of Ohio, he had presided at the convention which had renominated Lincoln. Now he had been paid off. He was the Postmaster General of the United States. The angry-faced Andrew Johnson addressed himself to his plate at the Kirkwood House. When he ate, a Negro servant always stood behind him. The Secretary of the Interior, John

P. Usher, a gray-faced man with poached eyes, sat at his front window looking through a lace curtain at the people passing by. What he was thinking about, no one knows. He was a friend of Lincoln's from the circuit-riding days. Soon, he would quit and go home.

On the north side of Lafayette Square, the suspicious Santa of the administration left for his office. This was Gideon Welles, Secretary of the Navy. He had full whiskers and puffed cheeks. At sixty-three, he was certain that Washington City was full of intrigue, and he kept a devastating diary of what the Cabinet members said and did. He disliked banks and seldom forgot either a name or a face.

Down in the Navy Yard section, David Herold sat on the edge of his bed and wondered what time it was. David was twenty-three, looked seventeen, and had the mentality of a boy of eleven. His nose curved like a scimitar and he had black hair and a manner of shifting from foot to foot when someone was talking.

He was unemployed now. Two years ago, he had had a job at Thompson's drugstore on Fifteenth at New York Avenue. Once he had delivered a bottle of castor oil to the White House and the President had asked him to charge it. This young man had three loves: John Wilkes Booth; hunting in southern Maryland; and practical jokes. His only resentment was that fate had given him seven sisters, the oldest of whom lectured him, the youngest of whom giggled when he tried to raise a mustache.

The man who slept late this morning was Lewis Paine. He was at Herndon House on Ninth Street, a block from Ford's Theatre. He was formidable, even in repose. Paine had no job, no ambition, no money, no girl. His only desire was to please Booth, whom he called "Cap." He placed no higher value on the lives of others than he did on his own, which was none.

Three and a half blocks north, Mrs. Mary E. Surratt was already cleaning up after breakfast. She was assisted by her daughter Anna, who was seventeen and who had secretly pasted Booth's picture behind a lithograph in her room, and by Susan Mahoney, an ex-slave. Breakfast was a detailed operation at 541 H Street, because this was the Surratt boardinghouse. It was here that the conspirators met.

Mrs. Surratt was a small woman, forty-five, a widow with a plain, unlovely face who parted her mousy hair in the middle and combed it back behind her neck. Her attitude was one of forced cheerfulness, as though she were being brave in the face of impending disaster.

8 a.m.

President Lincoln was a lean eater. His meals were not uniform, but they were nearly so. For breakfast, he usually ate one egg and drank one cup of coffee. For lunch, he ate one biscuit, drank one glass of milk, and ate one apple. At dinner, he sometimes drank hot soup, and almost always ate meat and potatoes. He would eat dessert if the dessert was homemade apple pie.

This morning, he wanted to listen to his grown son Robert. The boy had returned from a tour of duty with General Grant, and Lincoln wanted his son's firsthand opinions about Grant and the last days of the war. Mrs. Lincoln sat at the opposite end of the table, flanked by her sons Robert and Tad.

She was birdlike and happy this morning and, as Robert talked about the genius of Grant, she interrupted with admonitions to little Tad that he couldn't possibly play on the south grounds today because it was miserable out, and chilly too. Robert, as a joke, presented a picture of General Robert E. Lee to his father and the President, far from taking it as a joke, wiped his glasses on a napkin, studied the portrait a long time, and then said: "It is a good face. I am glad the war is over at last."

Mrs. Lincoln said that she had tickets for Grover's Theatre, which was staging a tremendous celebration tonight, but that she would rather see Laura Keene in *Our American Cousin* at Ford's. The President, listening to Robert, seemed to have little interest except to say that he would take care of it. Mother asked Robert if he would join them and he said that he was sorry, but that he had promised some friends that he would spend the evening with them. Tad said that nobody asked him to go. He was twelve and had a cleft palate and a sibilating lisp which kept him from regular attendance at school. This may be the reason that he was his father's favorite. The President spoiled him. Tad was the only person, highly placed or low, who could break in on a conference of state with no apology. Another reason may have been that the President had lost two sons. Edward Baker Lincoln died in 1850, not quite four years of age. In the same year, Mrs. Lincoln had given birth to William Wallace Lincoln. He died on February 20, 1862, of typhoid. At that time, the mother's grief was so severe that she was not permitted to attend the funeral and, for some time afterward, she had tried to communicate with Willie at spiritualistic séances.

Robert, the firstborn, was now twenty-one, a good-looking youth with brown hair plastered flat, and a mustache. He had graduated from Harvard University and he wanted to "join the colors," but Mrs. Lincoln was opposed to the idea, having lost enough sons to natural causes without risking a death in action. After much family discussion, the President did something that, for him, was mean. He asked General Grant to give the boy a commission and to place him on his personal staff, the inference being that he did not want Robert to be in danger.

Now the boy was home and he bubbled with stories about the siege of Petersburg and the gallantry of General Sheridan. His mother asked if he would not like to have the tickets to Grover's Theatre and he said he could use them, or give them to friends.

Mrs. Lincoln asked if the Grants would join them at the theater and the President said that he guessed they would, that Grant was a great hero and that the people were entitled to a look at him. His wife, who had a facility for cutting across conversational lines, asked if he would have time today for an afternoon drive. The President said that he did not know; that he would see. She said she wished that he would get out in the sun more often.

This was, at the moment, a happy family at breakfast. It wasn't always like this; in fact, it is no exaggeration to say that it was seldom like this. With the exception of Robert, it was a tragic group. Two of the boys were dead; the father would be dead by this time tomorrow; Tad would live six years; in ten years the mother would be certified as a "lunatic." In the family circle, there was often an aura of tension.

The mainspring of the family was not the President. It was Mary Todd Lincoln. She ruled by negation, and by fear. She was ten years younger than the President—he was fifty-six; she was forty-six. She was extravagant, economically and emotionally. Her inaugural dress of 1865 cost $2,000: In four months' time, she had bought three hundred pairs of gloves.

As a child, she was proud and haughty. Like John Wilkes Booth, she had an astonishing memory for long classical poems. She had a passion for sewing and her admiration in the early years went to her maternal grandmother, who ruled in chilly hauteur. When something was mildly funny, Mary laughed hysterically. When something was mildly sad, she wept hysterically.

When she came to the White House for the first time in 1861, she was short, plump, had a broad round forehead, light brown hair with iridescent lights of bronze, and a full figure. She wore her hair cleaved straight down the middle. When she stepped into the Blue Room for the first time, she said: "It's mine! My very own! At last it is mine!"

At once she changed the established order, fired the veteran White House steward, rearranged the furniture, put famous paintings in storage, hid heirlooms, ordered new décor, raged over food bills, accused servants of stealing, stalked through the corridors, not realizing that her taffeta whispered and warned servants that she was coming, refused to appear at state functions if she was piqued, excoriated Congress publicly for not giving the President enough money for state dinners, referred to General McClellan as "humbug" and called General Grant a "butcher."

She wanted to be addressed as "Madame President." The best the President could do was call her "Mother." Her funds were almost always low, or nonexistent. When bills for gowns were delivered to her, she often went into a frenzy of despair. At the back door of the White House, she berated butchers and grocers for their charges and she would clamor like the lid of a simmering saucepan until her eyes bugged and her voice failed. Afterward, she would sit shaking and palsied, exhausted.

One tradesman reached a point where he complained to Mr. Lincoln. The President is said to have eyed him sadly, put a strong hand on his shoulder, and said: "You ought to be able to stand, for fifteen minutes, what I have stood for fifteen years." This was in the early years, and once, at home in Springfield, her nagging reached a point where legend says that Lincoln lost control, laid violent hands on her, shoved her out the front door, and said: "You make the house intolerable! Damn you, get out!"

After that, he surrendered. Never again did he fight for reason. He did things against his will because they would please her. When he was in his office, and heard the first far-off peal of thunder, he ran from behind his desk to her side, because he knew that she would be terrified. A tree surgeon once approached him, perplexed, and said that Mrs. Lincoln insisted that a particularly fine White House tree be cut down. Lincoln didn't ask, "What tree?" He said: "Then, for God's sake, let it be cut down."

On the evenings of state balls, the President, dressed in somber tails and pulling on white gloves, always stopped in his wife's dressing room and said, almost cheerily: "Mother, which women may I speak to tonight?"

The mulatto seamstress, Mrs. Elizabeth Keckley, said that Mrs. Lincoln would tell him.

As she grew older, her rages became more violent, more enduring, and with them came dark hallucinations. Once he led her gently to a White House window and pointed to a big white building in the morning sun. "Mother," he said gently, "if you don't stop it you will spend the rest of your days there." The white building was the insane asylum.

She had terrifying dreams and often, when she awakened crying, he would hear her and get out of his bed and hurry to her and put his arms around her and comfort her. A friend asked what he thought of marriage, and Lincoln said quietly: "My father always said, when you make a bad bargain, hug it the tighter."

Mary Todd Lincoln felt that she was a shrewd politician. There were few vacancies of offices for which she did not have a candidate. Sometimes, to appease her, the President made the appointment. At other times, he ignored her recommendations. When he turned her down on a Secretary of State, and appointed Mr. Seward, she said, "That dirty abolition sneak!"

In the last campaign, she told Lizzie Keckley that Lincoln had to be reelected because she owed bills totaling at least $27,000 and there was no other way in which they could be paid.

She was generous too. When gifts of fruits and wines and liquors reached the White House, she loaded them in a carriage and drove out to the Soldiers' Home on Seventh Street and gave them to the wounded. She was never too weary to make the trip out, and nothing could make her hurry away from the bedsides. On afternoons when she noted that her husband looked tired, she often dropped her own plans and suggested that he come with her for an afternoon drive. At other times, unknown to him, she invited old Illinois friends to breakfast so that his mood might be brightened.

They were still chatting around the breakfast table when, far to the south, Jefferson Davis penned a note, in Greensboro, North Carolina, to his wife. He had no taste for breakfast this morning. The Confederacy was crushed and dead. He did not know whether the North would demand his life. And so his lean face was hard and expressionless as he penned:

Dear Winnie,
  I will come to you if I can. Everything is dark. You
should prepare for the worst. . . . My love to the children.

The letter had to be written early. Later in the day, there would be no time, because the President of the Confederate States of America had scheduled a conference with the leader of the last complete army in the field: General Joseph Johnston. And the bitterness that each of these men felt for the other would, on this day at least, be buried and Johnston would say: "My views are, sir, that our people are tired of the war, feel themselves whipped, and will not fight." At which point Jefferson Davis would write another letter, one to General William Tecumseh Sherman, asking for terms.

Across town from the White House, Noah Brooks was also writing. He was a newspaperman and, quite regularly, he wrote articles for newspapers in distant cities. The weakness in what Mr. Brooks wrote was the knowledge that he was a particular friend to the President. So particular, in fact, that in this coming June, Noah Brooks was scheduled to replace John G. Nicolay as one of Lincoln's secretaries.

Across the top of the first page, he wrote "News letter" in longhand. Underneath, he wrote: "April 14, 1865." The young man thought for a while, pen off paper, eyes vacantly staring through the window of his room, then the pen began to skate lightly over the paper, describing the elegant maneuvers which made the words which made the sense of his thinking. He had heard the President's speech of Tuesday, in which Lincoln had noted that elections were held in Louisiana and, as a sop to the radicals of his own party who wanted a harsh peace for the conquered South, he had said that he hoped the vote would be given to the intelligent Negro and the Negro who had fought in uniform.

"The radicals," Mr. Brooks wrote, "are as virulent and bitter as ever." He named some names, but he mitigated the sting of the high-sounding names by adding that the President's enemies "form but an inconsiderable portion of the great mass of the loyal people." The people, Brooks found, "have an implicit and truthful faith in Lincoln, which is almost unreasonable and unreasoning."

The reporter was exaggerating. The people of the South had no faith in Lincoln. The people of the North felt, in the main, that he was a stumbling, homely man whose "wrong" guesses, comically enough, were always justified in the end. "Old Abe will come out all right," they said, and, in this, one can read the chuckling affection one would have for a backward neighbor who always bests the pompous banker. Lincoln was lonely in a sense, but his isolation, such as it was, was created more by the politicians than by the people.

What Mr. Brooks was trying to say was that, on this Good Friday, the Republican Party was cracking under the President's feet in much the same manner that it had split under him four years before. The Butlers, the Wades, the Greeleys and the Sumners, added to the Democrats who despised him, would undo Lincoln. To them, the President stood for a soft peace; a let's-bind-the-wounds-and-get-back-to-work peace. Many of the ranking Republicans in the Senate and House wanted peace with a whip. The South was on her knees and, to them, it was not enough. They wanted to see her bleed. That proud trembling chin must be brought down.

The President got up from the breakfast table and said that he must be off to work. He walked back to his office, nodding to those who waited outside, and sat down at his big desk.

His office was a big square one in the southeast corner of the White House. In the center was a round oak table where Cabinet meetings were held. It was covered with a heavy green tasseled cloth. Around the room were chairs and two horsehair sofas. At the south end of the room—where he had sat reading earlier—were his pigeonhole desk and the small window table and a worn old chair. Along the north wall, near the entrance to the office, was a small door inside which was a basin, a mirror, a wall-bracket gas lamp, some soap, a towel and a comb.

Books in the room were few: a Bible, the Statutes of the United States of America, and a complete set of Shakespeare's works. Over a mantel hung a black-and-white engraving of President Andrew Jackson. Behind Mr. Lincoln's chair was a velvet bell cord for summoning secretaries. A soldier outside the office door brought the cards of guests.

The President sat, and, before admitting the first visitor, scanned the morning newspapers. He often said that he seldom read them.

At the same time, Mrs. Lincoln and her sons left the breakfast table and walked into the Red Room for a chat. The Lincolns had found this room cozier and more to their liking than any of the big rooms in the house, and it had become a family sitting room.

Upstairs, a maid began the task of straightening out the President's bedroom. The bed was low and large and the sheets were smoothed and the comforter was folded across the bottom. Two extra large pillows were set standing at the head. Over the head of the bed was a dark velvet canopy with lace side panels.

A plain cane chair beside the bed served for removing shoes and socks. A big brass ceiling fixture fed gas to white globes, although one was tapped with a hose which fed a small reading lamp on a round table. Two of the chairs near this table were cane; a third one was upholstered and had an anti-macassar. A big chair was placed with its back to the western light. Two rosewood folding doors connected with Mrs. Lincoln's bedroom.

The Days Before

The President

To see this one day clearly, it is necessary to see the President—and later, John Wilkes Booth—in the weeks prior to the event. Some of what happened on April 14, 1865, had earlier motivation. Some did not. Still, a certain pattern of events can be seen, in retrospect, and this pattern tends to increase, rather than diminish, the shame of the United States Government on April 14.

It seems, from the testimony of many witnesses after the event, that the government in early 1865 had two main conversational functions: killing the Confederacy, and keeping Lincoln alive. When the officials weren't talking about victory, and the means to victory, they were talking about the possibility of assassination. They talked about it, they worried about it and they counter-plotted against it. However, they were assuming that an assassination plot would involve the Confederate States of America *versus* the United States of America, and it seems not to have occurred to any ranking official that it might be a lonely band of fanatics *versus* the United States Government.

The newspapers of late 1864 and 1865 published dramatic and fretful stories of the narrow escapes of President Lincoln. Stanton's bureau of spies were uncovering plots in Richmond and in Washington almost weekly. The newspapers of the North, with or without the cooperation of the Secretary of War, published stories of the narrow escapes of the President. In the main, these plots probably did not exist, but, as the War Between the States moved toward its close, the stories made the people conscious of assassination and pressure was brought to bear on Stanton and on the President to be more and more careful.

In the early part of 1865, four members of the Washington metropolitan police force were appointed to guard the President. Two were on duty daily from 8 A.M. until 4 P.M. A third came on duty at 4 P.M. and remained at Mr. Lincoln's side until midnight. The fourth man arrived at midnight, and sat in the hall outside Mr. Lincoln's bedroom until relieved at 8 A.M. These men were not in uniform. Each had been trained in the use of the .38 pistol. Their specific order was to remain within a few feet of the President at all times and, in public, to look for faces they could not vouch for.

At about the same time, Stanton, not satisfied that four guards were enough, selected a troop of Ohio light cavalry, men who were mounted on fine black horses, and ordered them to act as presidential escort any time Lincoln left the Executive Mansion. This troop was quartered next door to the White House and, around the clock, they always had four horses saddled and bridled.

The first reaction to all of this was relief on the part of the Cabinet, irritation on the part of Mr. Lincoln. He said that Stanton was going too far. Later, he was amused by all the fuss and furor whenever he left the White House for an afternoon drive and, on some days, he made a game of trying to evade the cavalry escort. He did not try to "lose" his four policemen and, in time, cultivated them and sometimes confided personal opinions to them.

Withal, everyone worried about assassination and no one believed it would happen. Except one. Ward Hill Lamon not only feared it—he was certain that it would happen. In 1864, this fear overpowered him so much that, in stretches, he slept in the hall outside Lincoln's bedroom. Assassination, to him, was an *idée fixe*.

Lamon was a chunky-chested man with brown wavy hair and beard. He was the United States Marshal for the District of Columbia. He and Lincoln were old and dear friends, close enough to quarrel. On one occasion, when the President and two guests evaded the guards and attended the theater, Lamon, at 1:30 A.M., wrote in bitter sarcasm to his friend that neither of his guests "could defend himself against an assault by any able-bodied woman in this city."

The President trusted Lamon as he trusted few men, but he could not share his fears because Mr. Lincoln's philosophy was that he could be killed at any time by anyone who was willing to give his own life in return. Now and then, the President discussed a violent death, and, in this, his attitude was one of sadness and resignation rather than fright.

Still, the days of March were shiny with victory and short-term promise. The dusty banners of the Union snapped southward out of Fredericksburg, westward out of Old Point, northward out of Savannah, eastward out of Lynchburg. The noose tightened, hour by hour. The city to watch was Petersburg. When that fell, the final kill would occur at Richmond.

The South fought with valiance and empty bellies. The remainders of the great commands had pride and a little ammunition. The weaker Lee grew, the more craftily he planned. His men fought as though they could still win, and, man for man, perhaps they were the better soldiers.

General Ulysses S. Grant, a modest tenacious man who understood the value of numbers, repeatedly curled the whip of his Army of the Potomac around Lee's legs, and waited for his adversary to ask for mercy. Each day, Lee was a little bit weaker than yesterday. Each day, Grant snapped the whip a little harder.

The people of the North understood this. The news of victory in battle did not excite them, because they had waited an eternal four years for this. What excited them was the feeling that the war was in the twilight stage and soon it must end. They sensed this, and hungered for it.

They celebrated Lincoln's second inaugural as they had not celebrated the secretive first one. They came into Washington on excursion trains and in stagecoaches and carryalls and farm wagons, and they jammed the hotels and the boarding houses and the outlying farms until mattresses were stretched in the corridors of Willard's Hotel and the Baltimore & Ohio Railroad begged the people either to stay home, or to stop off at the City of Baltimore before inauguration day.

On the night before the inauguration, Lincoln worked in his office—perhaps on his "With Malice Toward None" speech, and Vice President Andrew Johnson, a one-time tailor and professional common man from Tennessee, attended a party tendered by Colonel John Forney, clerk to the United States Senate. The Vice President had been ill. He was also a poor drinker. That night, he got drunk. Sick drunk.

The day of March 4 was rainy. The mud on Pennsylvania Avenue was viscous. Below the inauguration platform, on the east plaza of the Capitol, the crowd looked like a vast bed of dark mushrooms. Under the umbrellas, they were jubilant. In the rain, the women preened in their best gowns and the men set up yells for Lincoln, Grant, Sheridan, Meade, Admiral Porter, and even McClellan.

In the Senate gallery, Mrs. Lincoln was attended by the courtly Senator Anthony. Admiral Farragut and General Hooker arrived at almost the same moment. The diplomatic corps, in dazzling uniforms and cocked hats, came to rest in the gallery as the justices of the United States Supreme Court came into the well below. The President sat in the front row on a low seat, his knees high, his ancient face lighted briefly by recognition and a slow nod of the head. Seward sat at his left, followed by Stanton, Welles, Speed and Dennison.

The Vice President would be sworn in here. Afterward, Mr. Lincoln would be sworn in, for his second term, outside. Johnson arrived on the arm of the outgoing Vice President, Hannibal Hamlin. Mr. Hamlin sang a short swan song, full of the rich sentimentality of the times, and introduced Andrew Johnson, who stood, red-faced and nervous and intoxicated, and who, in a few minutes, disgraced himself.

Mr. Johnson's theme was "the people." His effort was to demean himself before these high personages but, having done it, he did it again. And again.

" . . . for today," he shouted belligerently, as his audience glanced at neighboring faces, "one who claims no high descent, one who comes from the ranks of the people, stands, by the choice of a free constituency, in the second place in this government. . . .

"You, Mr. Secretary Seward, Mr. Secretary Stanton, the Secretary of the Navy, and the others who are your associates, you know that you have my respect and my confidence— derive not your greatness and your power alone from President Lincoln. . . . Humble as I am, plebeian as I may be deemed, permit me in the presence of this brilliant assemblage to enunciate the truth that courts and cabinets, the President and his advisers, derive their power and their greatness from the people."

The "brilliant assemblage" was shocked. The little man was drunk, obviously drunk. No one knew that Johnson, depressed by a hangover and his consciousness of his own tailor shop peasantry, had pleaded nervousness and had asked Mr. Hamlin for a drink. He felt sick, he said. So Hamlin had got a bottle, and handed it to Mr. Johnson, who was not in any stronger condition to fight the effects of whiskey today than he had been last night.

Hamlin pulled at Johnson's coattails. Forney whispered loudly to please sit down. Out front, the faces were frozen.

Lincoln alone looked sad and composed. Stanton was pop-eyed. Speed closed his eyes and held a hand over them. Justice Nelson's mouth hung open in horror. The gallery whispered. At last, after what might be called a Tennessee stump speech, Andrew Johnson took the Bible in both hands, kissed it loudly, and said: "I kiss this book in the face of my nation of the United States."

There was disgust in the crowd, a disgust resistant to wear, and Johnson would feel it the rest of his days. The whole party moved out to the plaza and President Lincoln was sworn in. His address amounted to less than four pages of copy:

"Both parties deprecated war; but one of them would make war rather than let the nation survive; and the other would accept war rather than let it perish. And the war came. . . .

"Both read the same Bible, and pray to the same God; and each invoked His aid against the other. It may seem strange that any man should dare to ask a just God's assistance in wringing their bread from the sweat of other men's faces; but, let us judge not that we be not judged. The prayers of both could not be answered; that of neither has been answered fully. . . ."

Within five minutes, he had closed his address with the words: "With malice toward none; with charity for all; with firmness in the right, as God gives us to see the right, let us strive on to finish the work we are in; to bind up the nation's wounds; to care for him who shall have borne the battle, and for his widow, and his orphan—to do all which may achieve and cherish a just, and a lasting peace, among ourselves, and with all nations."

He turned to leave. As he saw the faces of his distinguished friends nodding and applauding behind him, he smiled. Is it too much to say that, close by, in the inauguration stand, his eyes may have paused for a part of a second on a handsome young stranger—John Wilkes Booth?

Mr. Lincoln rode back to the White House with Tad at his side. The parade was a big one and a noisy one. At forts, the guns boomed, on the Avenue mounted patrols sat their horses at every crossing. The bands played. The President gravely raised his top hat to the people.

The handsome stranger left the inauguration stand with Walter Burton, the night clerk at the National Hotel. They walked back to the hotel bar for a celebratory drink. Booth had little to say, except that he had got his inauguration stand passes from the daughter of Senator Hale of New Hampshire.

The President's speech got perfunctory attention from the press, which labeled it "conciliatory" toward the South. The big news was not published; it was whispered. Andy Johnson had been drunk. Within two days, it had been whispered across the final dining-room table, the last bar, the ultimate alley, and Johnson, in anguish over it, left Washington for the home of Frank Blair in Silver Spring, Maryland. The gossipy United States Senate, which wanted to give the matter attention without being called guilty of bad manners, permitted a resolution to be offered prohibiting the sale of spirituous liquors in Capitol restaurants. In caucus, some Senators said that Johnson should resign.

Lincoln had seen this thing, and now he heard about it from all sides. He put it in its proper perspective when he said to Secretary McCulloch: "Oh well. Don't you bother about Andy Johnson's drinking. He made a bad slip the other day, but I have known Andy a great many years and he is no drunkard."

He defended Johnson, and yet the President was irritated because, after that, he would not see the Vice President although Johnson remained in and around Washington, waiting for an interview. Mr. Lincoln might have been hospitable and invited him to a Cabinet meeting or two, but he did not. The next time that the President expressed a desire to see "Andy" was on the one unique day when a clairvoyant chief executive would feel the need of a Vice President—April 14, 1865—*the* day.

Mr. Lincoln was thirty-five pounds underweight. He walked like a man whose feet hurt. Now and then, in the spring of 1865, he permitted a coachman to assist him in or out of his carriage. Lincoln was fifty-six; he looked old and sick. He had fought as hard as any soldier in the field to reunite the states. Now, in the closing days of the war, his spirit seemed to flag. The Surgeon General, Doctor Barnes, was worried about a nervous breakdown. The official family began to speculate, for the first time, about what would happen to them and to the nation if he died.

For a while, everyone including Mrs. Lincoln became solicitous of his health and his time. The police guards cleared the upstairs corridor of office seekers and favor seekers. His secretaries tried to hold the appointment calendar down, and his wife tried to coax him to take afternoon drives on sunny days. The attention was so pointed that even the President noticed it.

On March 14—a Tuesday—Mr. Lincoln tried to arise from his bed and fell back. He could not summon the strength to get on his feet and Mrs. Lincoln, called from her bedroom, sent for Dr. Robert K. Stone, the family physician. He examined the President and came out of the bedroom announcing that the case was one of "exhaustion, complete exhaustion."

Three hours later, the President held a Cabinet meeting in the bedroom. Word went out that his illness was influenza, and it was so reported in the press. By Wednesday morning, he was out of bed and in his office. He looked jaundiced and sick, and no appointments were made that day, but he worked at his desk. In the afternoon, Mrs. Lincoln told him that she could be made to be very happy if he felt strong enough to attend

Grover's Theatre with her. The President consented. They saw the German Opera Company perform *The Magic Flute.*

The newspapers began to worry about the state of Lincoln's health. *The National Republican* proclaimed that he was physically exhausted as a result of prosecuting the war to a successful conclusion and that, in addition, he was badgered by swarms of office seekers who infested the halls of the Executive Mansion. The writer recommended that these people be driven from Washington at once, or else the nation would run the risk of a presidential breakdown. The New York *Tribune* went a step further. The President's energies, said a lead editorial, would have to be spared if Lincoln was to live through his second term. Unless something is done promptly, the Union will mourn a dead President "killed by the greed and impudence of bores."

In the third week of March, General Grant felt that the end of the war was a matter of time—a few weeks at most. By army telegraph, he invited the President to come down to City Point, Virginia, to see the end. Petersburg was under siege and Grant was encamped before it. Philip Sheridan had traversed the Shenandoah Valley, destroying and requisitioning as he went, and Lee no longer had enough food for horses and men. William Tecumseh Sherman had finished with Savannah and swung north toward Virginia to hammer Lee and Johnston against the anvil of the Army of the Potomac. Sheridan rejoined Grant, and his mobile force was used, like an oversized hound dog, to keep the Confederate rabbit from running out of the pen.

Lincoln left Washington aboard the steamer *River Queen* on Thursday, March 23. With him were Mrs. Lincoln, her maid, Tad, bodyguard William H. Crook, and Captain Charles B. Penrose. The boat sailed at 1 P.M. and, as the heavy lines were cast off, the President seemed to cast off the heavy shackles of office. He romped in the forward salon with Tad, and took him up to the wheelhouse to show him how a steamer operated. He radiated relief and suppressed excitement and, to Mrs. Lincoln, seemed full of mischief. The First Lady brought several trunks of finery with her and told the President that she had heard that General Grant had ordered all the generals' ladies to the rear and she supposed that she would be the only female at the front. Mr. Lincoln said that he had heard no such thing.

The *River Queen* made port at City Point the following day. The quayside was choked with the appurtenances of war—guns, cannon, barrels of pitch, ammunition, lumber, foodstuffs and a private railroad for the Army of the Potomac. In the President's honor, the ranking generals were present, and came aboard for two conferences within a few days. Grant, Sherman and Sheridan were there (Meade was at the front), and Grant, assisted by the other two, spread the war maps in the dining salon and rendered a firsthand account of the final operations of the war.

Grant told Lincoln that the war had now been reduced to a matter of arithmetic. Lee and Johnston were losing a regiment a day through desertion, illness and wounds. They had no manpower to replace such losses. The longer Lee held out, the weaker he became. If he elected to fight battles, he would hasten the end—even if he won the battles—because his losses would increase. The end, the President was told, would come within a month.

On the maps, Grant showed the President that the one move left to the Army of Northern Virginia was to abandon Richmond and Petersburg and try to consolidate forces with Johnston on the Virginia-North Carolina line. Grant proposed to devote his time to preventing that move.

Two ladies bumped over a corduroy road on Saturday. This was an army road and, in a circuitous way, it led to the front.

One of the ladies was Mrs. Lincoln. The other was Julia Dent Grant. They rode in an army ambulance and both held on to the overhead hoops to keep from falling. They sat on a crossboard seat. Directly in front were an army driver and General Adam Badeau, who had been ordered to escort the distinguished ladies to a troop review.

Mrs. Lincoln chattered happily as the wagon swayed over the pine logs. Even here, in a forest of saplings and old undergrowth, she was dressed lavishly. Mrs. Grant did not carry the conversation. She was a plain, almost homely woman with a long nose, and what she smelled in the presence of the First Lady of the Land was trouble, Julia Grant could never explain it, but she always felt nervous in the presence of Mrs. Lincoln. So, as the mules walked, she merely nodded yes, and yes and yes.

On her side, Mrs. Lincoln was careful too. She now knew that General Grant had indeed issued an order for all wives of general officers to go to the rear, but he had exempted his own wife. Mrs. Lincoln did not mention it, although it may have been close to her tongue.

For his part, General Badeau, a tactful man and a keen intellect, filled the gaps in the conversation by explaining that it was Crawford's division which would pass in review, and that the salute would be taken by General George Meade, Commander of the Army of the Potomac. The ladies paid little attention. However, when he moved blithely on to the subject of Grant ordering women out of the area, he got complete attention. The wagon jarred and swayed as he explained, with humor, that when such an order came through, the men in the ranks regarded it as a sure sign that a battle was impending. This time, he said, General Grant had permitted no exemptions except, of course, Mrs. Grant and Mrs. Charles Griffin. Mrs. Griffin remained at the front, he said, because she had a special permit from the President.

Mrs. Lincoln's rages were always almost instantaneous. This time, she almost rose from her seat. "What do you mean by that, sir?" she snapped. "Do you mean to say that she saw the President alone?" Mrs. Grant turned to her in alarm. General Badeau looked over his shoulder at the stricken woman. "Do you know that I never allow the President to see any woman alone?"

The pitch of the voice rose higher. Julia Grant looked ill. Badeau swung all the way around and tried to smile reassuringly. "That," said Mrs. Lincoln, "is a very equivocal smile, sir! Let me out of this carriage at once!" She started to clamber toward the canvas side of the wagon. "I will ask the President if he saw that woman alone."

Mrs. Grant tried to soothe Mrs. Lincoln. The First Lady was now livid. She wasn't listening. The general, who seemed confused, apologized even though he wasn't certain of the offense. Mrs. Lincoln ordered the ambulance stopped at once. When the mules continued their slow pace, she reached past the driver's shoulder and tried to yank the reins. The general's wife, almost in tears, begged Mrs. Lincoln to please sit down. Just sit.

The First Lady sat. She was silent, and her face twitched. The ambulance continued on the road. Nobody spoke. When it arrived at the parade grounds, General Meade came to the steps at the back of the ambulance and assisted Mrs. Lincoln.

After the review was over, Mrs. Lincoln returned to the ambulance and got in and stared at the back of Adam Badeau's head. "General Meade is a gentleman, sir," she said. "He says it was not the President who gave Mrs. Griffin the permit, but the Secretary of War."

At City Point, Mrs. Grant spoke to General Badeau in private and begged him "never to mention this distressing and mortifying affair again."

The month of March closed with minor chords. In Washington City, the headquarters of General C. C. Augur, at 151/2

Street and Pennsylvania Avenue, was badly damaged by fire and Augur had to move most of his staff to Fourteenth Street. It meant that the White House would not have the protection of extra guards next door. In the War Department, Stanton and Eckert and Bates listened to the snap of the telegraph keys and hoped for momentous news, but the best they got was signed "Lincoln": "There has been much hard fighting this morning. . . . Our troops, after being driven back on the Boyd-ton plank road, turned and drove the enemy in turn and took the White Oak road. . . . There have been four flags captured today."

The massive crescendos began to be heard in the opening days of April and some of the counterpoint was lost in the bedlam of sound. For instance, on April 2, the Confederacy fell. No one in the North knew it, and no newspaper carried the news, but, shortly after church services in Richmond, Jef-ferson Davis and his Cabinet fled the city. That was the end, even though Lee and Johnston still fought in the East. There was no roll of covered drums, no ceremony, perhaps few tears.

On this Sunday, the Confederacy died and there was no longer an amalgamation of seceded states. An idea born of pride was gone, and had taken its place in the pages of history. After four years of battling a brother who was bigger, stronger, better fed, better armed, the South was whipped.

The President, still at the front, did not know it. His daily dispatch to Stanton said: "All going finely. Parke, Wright, and Ord, extending from the Appomattox to Hatcher's Run, have all broken through the enemy's intrenched lines, taking some forts, guns and prisoners. Sheridan, with his own cavalry, Fifth Corps, and part of the Second, is coming in from the west on the enemy's flank."

The next day, everybody knew the news. It started when Lincoln sent a coded message to the Secretary of War an-nouncing the portentous intelligence that the city of Petersburg had been evacuated by the Confederate Army and that Grant was "sure" Richmond too had been abandoned. Stanton had slept in his office and, when the code clerk had reduced the message to straight English, the Secretary of War was awakened and told the news.

He was barely savoring the exultation of it when a telegraph key started an insistent chatter in straight English:

"From Richmond," it began. Two army operators listened, in bug-eyed disbelief, then emitted a whoop and ordered a fifteen-year-old apprentice telegrapher, Willie Kettles, to copy the rest of it. The two operators threw up a front window and began to roar, in unison: "Richmond has fallen! Richmond has fallen! Richmond has fallen!" Citizens on the walk below looked up anxiously. The two yelled the louder. Drays on the cobbles, and carriages too, came to a stop. The faces below began to comprehend; they began to crease in attitudes of smiles, and relief, and sudden sadness and ecstasy. An aged man threw his hat down and jumped on it. A woman blessed herself. A wagon driver burst into tears and blew his nose.

The cry was taken up, and boys skittered down Pennsylvania Avenue in the cool yellow sunshine passing the word and bumping into people and the word began to spread quickly and wildly. When it reached the offices of the Washington *Star,* an editor ran out front and printed in chalk on a big blackboard:

GLORY!!!   HAIL COLUMBIA!!!
HALLELUJAH!!!   RICHMOND OURS!!!

In a public park, a battery of guns was limbering up when the word came. The officer in charge became so excited that he ordered an immediate salute of eight hundred rounds, three hundred for Petersburg and five hundred for Richmond. The cannonading was massive and, as it echoed across the

Navy Yard, an officer heard it and, not knowing the news, decided to fire one hundred rounds on a big Dahlgren gun on the chance that the news might be important. In an hour, the offices of Washington City were almost empty, and many of the stores were without clerks. Stranger hugged stranger and the taverns did a brisk morning business. The courts adjourned. Children skipped home from school. The banks closed. Church bells tolled in the hollows between mountainous crashes of artillery. Flags appeared before the homes of the loyal and the disloyal. Horse cars stopped running. An impromptu parade started on Sixth Street, the first of many. Negro families emerged from shacks shyly, like children hoping to be asked to a party. Unbidden orators stood on the several hotel steps, faces red, arms waving, but not a word was heard in the bedlam.

Mr. Stanton, surrendering to a rare moment of happiness, leaned from a War Department window and held up a hand for silence. He asked the crowd below to beg Providence "to teach us how to be humble in the midst of triumph." In the momentary vacuum, someone said that Richmond was burning, and the crowd roared: "Let 'er burn!" Willie Kettles was introduced from the telegraph window as the "man" who had received the auspicious message. Willie bowed gravely from the waist.

On E Street, two squadrons of cavalry met and, without orders, got into parade formation and in a moment a brigade of infantry fell in behind. As the parade moved, it grew. An hour later, the cavalrymen led it across the south grounds of the White House and they were surprised to find that they were being reviewed by General C. C. Augur.

This was going to go on, sporadically, for twelve days. It was the wildest celebration known to the young Republic and it would not end until the nation was plunged into deepest grief. On some days, it would flag a little, through surfeit or exhaustion, and then fresh news of victory would come and it would revive in Washington City and New York and Springfield and St. Paul and in the crossroad settlements across the country. The feeling in most minds was that two incredible things had happened: the war was over; we won it.

And yet it was not quite over. Lee's army was still in the field. It was a striking unit in being; it had its fighting units intact, its stores, its staff. It was dying in dignity, and no one in the Army of the Potomac was celebrating.

The President sent word to Stanton that he was about to sail upriver to take a look at burned-out Richmond, and the Secretary of War was beset with misgivings. Had he the power, Stanton would have placed the President under military detention to keep him out of Richmond. He knew that Mrs. Lincoln had returned to Washington yesterday, and would not return to her husband for a few days, but Stanton did not visit the White House on this Monday, April 3, to ask her to stop Lincoln. There is no record that he even visited to ask how she enjoyed her trip.

Instead, Stanton tried the direct approach. He sent a message to Lincoln:

"I congratulate you and the nation on the glorious news in your telegram just recd. Allow me respectfully to ask you to consider whether you ought to expose the nation to the consequences of any disaster to yourself in the pursuit of a treacherous and dangerous enemy like the rebel army. . . . Commanding Generals are in the line of their duty in running such risks. But is the political head of a nation in the same condition?"

The President read it, was comforted by the solicitude of his Secretary of War, and took a boat up the river to Richmond. When he disembarked at a riverbank on the edge of the Confederate capital, Admiral Porter was at his side. A small group of Negroes saw the overly tall man with the stovepipe hat scrambling up the bank and, when they saw the quizzical, sad smile on the thick lips, a few recognized him from half-remembered pictures. They began to shout and bow and a group gathered about him, a few kneeling to kiss his black shoes.

"Don't kneel to me," the President said sharply. "This is not right." He glanced around the circle of dark faces and saw the wonderment in them. "You must kneel to God only, and thank Him for the liberty you will hereafter enjoy. I am but God's humble instrument, but you may rest assured that, as long as I live, no one shall put a shackle to your limbs, and you shall have all the rights which God has given to every other free citizen of this Republic."

He had injected the somber note: *"As long as I live. . . ."*

Admiral Porter tried to push the Negroes away. They were as pliant as full wheat. They moved back when pushed. When the hand was removed, they returned. Negroes seemed to be coming from everywhere and the admiral looked around helplessly for help. There was none. The people pressed around the President and they sang and chanted. A few bold ones tried to touch the sleeve of his coat. Lincoln stepped forward, determined to see a part of Richmond, and the circle moved with him. Later, Lincoln was seen and rescued by a roving squadron of U.S. Cavalry. Both groups were equally surprised.

All of the nation's news came from the front in these final days. Washington City—normally the master maker of news—was, for the moment, a listening post. Congress had adjourned; many legislators had gone home to mend fences. The streets were full of men in uniform, men from Ohio and Vermont and Illinois and Delaware and Missouri who were on leave from one of the many camps in and around the city, and who wanted, before the war was done, to say that they had seen the new Capitol dome and the White House.

The most momentous event on Tuesday, April 4, was the arrival of the steamer *Thomas Powell* with three hundred wounded aboard. The most trivial news was that Mrs. Lincoln, preparing to return to City Point, Virginia, sat in the White House and wrote notes for two of the President's guards, detailing them to duty at the White House and, in effect, exempting them from being drafted into the Army. Both guards, John Parker and Joseph Sheldon, had been notified that they were being drafted, and both had asked Mrs. Lincoln for the note.

It was news of a happy sort that the State Department had ordered a grand illumination of all Federal buildings in the District of Columbia for this Tuesday night in celebration of the fall of the Confederate capital. All day, hundreds of workmen crawled along the façades of buildings carrying bunting. The Navy Department built a big model of a full-rigged ship and held it aloft in front of the building with piano wire. Over the front of the Treasury building, a gigantic ten-dollar bill could be seen. The main War Department building was hidden by hundreds of flags.

Stanton wanted his department to do a memorable thing and so, shortly after sunset, men were stationed in each window of the War Department's eleven buildings, armed with matches. At twilight, an army band crashed into "The Star-Spangled Banner" and, in an instant, the buildings swam in a pool of yellow flame. At the far end of Pennsylvania Avenue, for the first time, the Capitol was lighted from basement to dome by gas, and, across the front, in letters two stories high, blazed the message:

THIS IS THE LORD'S DOING; IT IS MARVELOUS IN OUR EYES.

The cheerful flicker of candles could be seen in almost every home on every street. Except at Mrs. Surratt's boardinghouse. Here the shades were drawn and the owner wept.

The celebrants were on all streets in rollicking bands. In front of the Patent Office, a crowd saw the Vice President and someone yelled "Speech! Speech!" Andrew Johnson, red of face and angry, said that the leader of the rebellion was Jefferson Davis, a West Point graduate who had plunged the sword given to him by his country into his mother's bosom. There were cries of "Hang him! Hang him!" and Johnson roared back: "Yes, hang him twenty times because treason is the greatest of the crimes!"

The glee of the people was reflected in the newspapers, which, in stories never wider than one column, made the news gladsome and official. The New York *Herald* was radiant in diminishing sizes of type:

GRANT

RICHMOND OURS

Weitzel Entered the Rebel Capital Yesterday Morning

MANY GUNS CAPTURED

Our Troops Received With Enthusiasm

An editorial, probably written by editor James Gordon Bennett, fed the dream of power politics to the people: "The end of our great Civil War is close at hand," it said. "It is very easy to see that with the return of peace, this country will be the greatest in the world. Midway between Europe and Asia, geographically, we shall hold the balance of power politically, commercially and financially. As our resources are developed we shall produce the gold, silver, iron, petroleum, corn and cotton for the use of all mankind. We are the center of the world, and we shall move everything by our immense central force. In creating this nation, Providence created the acme of strength and civilization. It is our manifest destiny to lead and rule all other nations."

On Wednesday, April 5, Mrs. Lincoln left Washington to rejoin her husband at the front. With her aboard the steamer were a party of friends. The war, within a few days, had taken on the aura of a sport, a hunt. Shipboard life was happily expectant. The steamer was barely past Indian Point when State Secretary William H. Seward, who had planned to join Lincoln to "sell" him on the idea of closing Southern ports to all but Northern traders, was out riding in his carriage. His matched blacks cut a corner too sharply and ran away. The front right wheel of the vehicle was smashed, and it screeched over the paving stones, acting as sled and brake at the same time. Seward pitched out and sustained a broken arm, a broken jaw, multiple contusions of face and head, and concussion of the brain. He was sixty-four.

A small thing occurred on April 6, a week and a day before *the* day. John Surratt, son of the boardinghouse widow, arrived in Montreal with dispatches from the Confederate Secretary of State, Mr. Judah Benjamin. For the next week, Surratt would be busy with Southern General Edwin G. Lee.

Now there was a period of quiet. From Thursday until Sunday, nothing of moment occurred except that General Robert E. Lee made a final, masterful attempt to haul his tired army southwestward to join General Johnston. Coming out of a small valley, his lead regiments saw horsemen on a ridge ahead.

General Philip Sheridan was calling check.

On Palm Sunday—April 9—the *River Queen* came upstream in the afternoon and docked with the President, Mrs. Lincoln, and a party of friends. Mr. Lincoln had heard about Seward's accident and, begging leave of the others, hurried on alone to his Secretary of State.

At the "Old Clubhouse"—the Seward home—Lincoln stood hat in hand in the lower hallway and listened to Frederick Seward retell the story of the accident and the grievous injuries. The President heard that Surgeon General Barnes had pronounced that, now that Seward had survived the initial shock, he would live. Lincoln walked up the two flights of stairs, and went to the bedroom at the front of the building on the left side. He tiptoed into the darkened room and, standing a moment, saw the secretary.

Seward was on the side of the bed away from the door. His face, the small part of it that was visible, was unrecognizable with swelling and discoloration. Bandages and dressings covered the entire head except for the purple eyes and the cruelly ripped mouth. Without moving the twice-broken jaw, he whispered:

"You are back from Richmond?"

"Yes," said the President, "and I think we are near the end, at last."

Without invitation, the President did something rare and impulsive. He sprawled, on his stomach, across the empty side of the bed, and he told his secretary all that had happened in Virginia in the past week. Lincoln was still talking, a half hour later, when he studied Seward's eyes and saw that he was sleeping. The President arose softly, in stages, and tiptoed from the room.

At 9 P.M. on that Palm Sunday night, Secretary of War Stanton was dozing on a downstairs couch in his home. An army messenger yanked the pull bell, found that it was broken, and drummed his fist on the front door. The secretary was awakened, and was given a dispatch:

Headquarters, Appomattox Ct. H. Va.
April 9, 1865   4:30 p.m.

> Hon. E. M. Stanton, Secretary of War, Washington
>   General Lee surrendered the Army of Northern
> Virginia this afternoon on terms proposed by myself.
> The accompanying additional correspondence will
> show the conditions fully.
>
> > *U.S. Grant,*
> > *Lieut.-General*

The iron man of the administration read it again. He was close to tears. While the messenger waited, he went to his desk, sat, and penned a reply:

> Thanks be to Almighty God for the great victory with
> which He has this day crowned you and the gallant
> army under your command. The thanks of the De-
> partment and of the government and of the people of
> all the United States, their reverence and honor, have
> been deserved and will be rendered to you and the
> brave and gallant officers of your army for all time.

It was too late in the evening for a celebration. Mr. Stanton did the next best thing. He dressed and hurried to the White House with the most momentous news of his career. The President was in the Red Room with Mrs. Lincoln and some friends, and Lincoln was standing with his back to the coal grate, flicking his coattails as the dispatch was read. It was greeted with stunned silence. Now that it had happened, it was beyond the capacity of these people to comprehend. All faces seemed blank; the expressions were almost the same as though the news had been bad.

Around this time—it may have been this night—Stanton

asked to see the President alone and handed a paper to him on which was written the War Secretary's resignation. Lincoln read it through, took the paper between his hands and tore it slowly, dropping the fragments into a basket, and placed his big hands on Stanton's shoulders.

"You cannot go," he said. "Reconstruction is more difficult and dangerous than construction or destruction. You have been our main reliance. You must help us through the final act. The bag is filled. It must be tied and tied securely. Some knots slip. Yours do not. You understand the situation better than anyone else, and it is my wish and the country's that you remain."

Washington City, tired and hungover from almost a week of celebrating, awakened on the morning of Monday, April 10, to the crashing of cannon. The people listened, and wondered what further good news was possible. Lee, they learned, had surrendered to Grant. The celebrating started all over again.

A big battery was firing in Massachusetts Square, near Scott Circle and, between basso blasts, the treble tinkle of window glass could be heard. The morning newspapers, hawked up and down the streets, told of the dramatic meeting between the generals, how Grant had permitted the Southern officers to be paroled and to retain their sidearms, and of how he permitted the defeated army to keep its mules and horses for plowing old ground. The editorial pages speculated that, with Lee out of the way, Joseph Johnston commanded the only sizable force left to the Confederate states and, caught between Grant and Sherman, it must capitulate within a few days.

That afternoon, the flags in the capital drooped in rain. People huddled before the White House in damp expectancy. Now and then, a cry of "Speech!" went up. The President sent word out that, because of his recent trip, he was behind in his work and he advised the people to disperse. One of the things he did on this bleak Monday was to sit for Gardner, the photographer. While the pictures were being taken, Tad frolicked around the room, bouncing on and off his father's lap, distracting Mr. Lincoln to the point that, for the first time, he smiled faintly in a picture.

Twice, the President went to the front windows of the White House, pulled back the curtain, and waved to the crowd below. He was waving when he saw Tad run out on the porch with a captured Rebel flag and race up and down in the dampness, trying to make the banner snap in the breeze. The crowd laughed when it saw the President of the United States, slightly harassed and embarrassed, come out to retrieve his son.

There was no way that the President could get back inside gracefully without saying something, and so, informally, he turned to the crowd, hanging on to Tad, and said that he supposed there would be some formal celebration, and that he would save his words for that occasion. There was scattered applause. The Navy Yard band was standing under the eaves and Lincoln asked the leader to please play a song for the people; "Dixie," he thought, would be appropriate because it could now be considered the lawful property of the United States.

When he returned to his desk, Lincoln found a message from the Department of State advising him that the formal celebration of Lee's surrender would be held on the evening of Tuesday, April 11 (tomorrow), and that there would be another grand illumination of the city with speeches, parades, etc.

The city was quiet on Tuesday. The people husbanded their strength for the evening and, shortly after 6 P.M. when the sun set, the festivities began. It was as spectacular as the earlier illumination and, when darkness had dusted the final alley, the Lee mansion in the hills across the river was aglow with lights, and freed slaves danced on the lawns before it, humming "The Year of Jubilee." The city swam in light and the people were as festive as though there had been no celebration like this in years.

The weather was warm and misty. The crowd before the White House had changed personnel two or three times and was now much larger. The people filled Pennsylvania Avenue and trampled the shrubs of the grounds. Small sections of the people were coned by the gas lamps and an observant reporter wrote: "There is something terrible in their enthusiasm."

A hanging mob had come to listen to a man of mercy.

The Marine band played marches. The crowd chanted "Lincoln! Lincoln!" The people undulated, those in back pressing forward, those in front holding the line. Two who pressed forward and managed to achieve a good position beside a tall tree were John Wilkes Booth and his friend Lewis Paine. Booth was impelled to hear the man he hated.

The people were becoming impatient when a French window was opened and the curtains pulled back on both sides. In silhouette, the President could be seen, waving both hands over his head. The cheers were frenzied. It was as though the people had not believed, until now, that this man could win. He waited gravely until he had near silence and then he unrolled a sheaf of foolscap and then rolled it in the opposite direction so that, as he held the pages, they would lie flat.

An arm appeared beside him, holding a lamp with a china shade. Mr. Lincoln adjusted his metal-rimmed spectacles and then began to read, so softly at first that the crowd heard but a whispering sound, then louder as he sensed the need for it until, after a few minutes, his voice was plain to all except those on the far side of the street.

They listened for exultation, and there was none. They strained for eloquence, and there was none. They waited patiently for vengeance, and there was none.

The President talked about Reconstruction. He talked soberly about postwar problems, as he saw them. He told them about the voting situation in Louisiana, where the lists were down from forty thousand to twelve thousand, arithmetic which only proved that Southerners would stay home from Yankee-sponsored elections. To cure this, Lincoln prescribed strong medicine.

"It is also unsatisfactory to some," he said slowly, "that the elective franchise is not given to the colored man." The crowd was quiet. "I would myself prefer that it were now conferred on the very intelligent, and on those who served our cause as soldiers. . . ."

John Wilkes Booth sucked in a long breath. He tapped Lewis Paine on the arm. "That's the last speech he will ever make," the actor said. The two men edged out of the crowd.

Lincoln finished his talk and the applause was restrained and respectful. He bowed and stepped back from the window. The second speaker was Senator James Harlan of Iowa, now Secretary-designate of the Department of the Interior. One day in the future, his daughter would marry Robert Lincoln.

Mr. Harlan had excellent intentions, but he did not know that a good speaker never asks an explosive mob a question.

"What," he said with arms outstretched, with silvery syllables echoing in the trees, "shall be done with these brethren of ours?"

As one, the crowd roared, "Hang 'em!"

The Senator smiled in the face of thunder and said that, after all, the President might exercise the power to pardon.

"Never!" the crowd screamed.

The Senator tried to educate and inform by suggesting that the great mass of Southern people were not guilty. He got silence. The Senator was not equal to further effort. He finished haltingly by proclaiming that he, for one, was willing to trust the future to the President of the United States.

He left the window and the people gave him an enthusiastic hand. The Marine band struck up "The Battle Hymn of the Republic" and, in a soft drizzle, the crowd broke up.

No one had, on this night of victory, counted the dead. The United States would never officially count the Confederate dead, would never even keep records of the Confederate wounded. Still, the North paid more in blood and treasure than the South. About 110,000 men, largely young and fair, died in battle or died of wounds. About a quarter of a million more died of diseases attributable to war. The South's losses, in battle and by disease, were about 133,000. Both sides paid in dead a little more than one and a half percent of the population of 31,000,000 people.

Inside the Executive Mansion, Mr. and Mrs. Lincoln entertained a few friends. In the Red Room, he sat beside her on a sofa and listened to her birdy chatter. At ten, tea and cakes were served and, shortly afterward, the friends began to make their adieux. That is, all except Ward Hill Lamon, Senator Harlan and his daughter, and one or two others. The dominant emotion seemed to be relief rather than happiness. It was difficult to talk in an evening of no tensions.

To make conversation, Mrs. Lincoln said that, in the midst of joy, her husband's face looked long and solemn. The President said that his mind had been heavy. The faces turned toward him.

"It seems strange," he said slowly, as though feeling for the words, "how much there is in the Bible about dreams. There are, I think, some sixteen chapters in the Old Testament and four or five in the New in which dreams are mentioned; and there are many other passages scattered throughout the book which refer to visions. If we believe the Bible, we must accept the fact that, in the old days, God and his angels came to men in their sleep and made themselves known in dreams."

Mr. Lincoln studied the suddenly solemn faces of his friends. He sat forward, elbows on knees, the veined hands describing small gestures.

"Nowadays," he said apologetically, "dreams are regarded as very foolish, and are seldom told, except by old women and by young men and maidens in love."

Mrs. Lincoln looked worried. "Why?" she said. "Do you believe in dreams?"

"I can't say that I do," he said, hedging against the nightmares she had suffered for many years, "but I had one the other night which has haunted me ever since. After it occurred, the first time I opened the Bible, strange as it may appear, it was at the twenty-eighth chapter of Genesis, which relates the wonderful dream Jacob had. I turned to other passages, and seemed to encounter a dream or a vision wherever I looked. I kept on turning the leaves of the old book, and everywhere my eyes fell upon passages recording matters strangely in keeping with my own thoughts—supernatural visitations, dreams, visions, and so forth."

Mrs. Lincoln clutched her bosom. "You frighten me," she breathed. "What is the matter?"

At once the President tried to dismiss it. "I am afraid," he said, "that I have done wrong to mention the subject at all. But somehow, the thing has gotten possession of me, and, like Banquo's ghost, it will not down."

He tried to talk of other things. Mrs. Lincoln would not be put off. She asked about the dream. Mr. Lincoln's face settled again in melancholy and he agreed to tell about it.

"About ten days ago,* I retired very late. I had been waiting up for important dispatches. I could not have been long in bed when I fell into a slumber, for I was weary. I soon began

---

* It is commonly agreed that Lincoln is wrong about the time of the dream. Lamon, who remembered this dialogue almost word for word, believed that the dream occurred on March 19, the night after the Lincolns saw *Faust*.

to dream. There seemed to be a deathlike stillness about me. Then I heard subdued sobs, as if a number of people were weeping. I thought I left my bed and wandered downstairs.

"There the silence was broken by the same pitiful sobbing, but the mourners were invisible. I went from room to room. No living person was in sight, but the same mournful sounds of distress met me as I passed along. It was light in all the rooms; every object was familiar to me, but where were all the people who were grieving as if their hearts would break?

"I was puzzled and alarmed. What could be the meaning of all this? Determined to find the cause of a state of things so mysterious and so shocking, I kept on until I arrived in the East Room, which I entered. There I met with a sickening surprise. Before me was a catafalque, on which rested a corpse in funeral vestments. Around it were stationed soldiers who were acting as guards; and there was a throng of people, some gazing mournfully upon the corpse, whose face was covered, others weeping pitifully.

"'Who is dead in the White House?' I demanded of one of the soldiers.

"'The President,' was his answer. 'He was killed by an assassin.'

"Then came a loud burst of grief from the crowd, which awoke me from my dream. I slept no more that night, and, although it was only a dream, I have been strangely annoyed by it ever since."

Mr. Lincoln fell silent. The story was over. Ward Hill Lamon looked at the faces in the room. No one spoke. Mrs. Lincoln looked frightened.

"That is horrid," she said. "I wish you had not told it. I am glad I don't believe in dreams, or I should be in terror from this time forth."

The President smiled. "It was only a dream, Mother. Let us say no more about it, and try to forget it."

Senator Harlan arose to say good night. Secretary of the Interior Usher elected to stay a moment longer. So did Ward Hill Lamon. The President had asked Lamon, as a favor, to go to Richmond as his personal representative, and to see that certain anticipated complications at a state convention were smoothed. "Hill" had already agreed to go. Now, when the others had departed, and Mrs. Lincoln had said her good nights, Usher and Lamon tried to persuade the President not to go out anymore after nightfall. Ward Hill Lamon practically begged the President not to go out until he returned from Richmond.

"Usher," Mr. Lincoln said banteringly, "this boy is a mono-maniac on the subject of my safety. I can hear him, or hear of his being around, at all times of the night, to prevent somebody from murdering me. He thinks I shall be killed, and we think he is going crazy." He grasped Hill's shoulders in his big hands and shook gently. "What does anybody want to assassinate me for? If anyone wants to do so, he can do it any day or night, if he is ready to give his life for mine. It is nonsense."

The Secretary of the Interior shook his head in disagreement. "Mr. Lincoln," he said, "it is well to listen and give heed to Lamon. He is thrown among people that give him opportunities to know more about such matters than we can know."

Lamon brought up the subject of the dream, and the President chided him, saying: "Don't you see how it will turn out? In this dream it was not me but some other fellow that was killed. It seems that this ghostly assassin tried his hand on someone else." Mr. Lincoln was trying hard to laugh. His friends stared at him. "And that reminds me," he said, "of an old farmer in Illinois whose family was made sick by eating greens. Some poisonous herb had got into the mess, and members of the family were in danger of dying. There was a half-witted boy in the family called Jake, and always afterward when they had greens the old man would say: 'Now, afore we risk these greens, let's try them on Jake. If he stands them, we're all right.'

"Just so with me. As long as this imaginary assassin continues to exercise himself on others, I can stand it." The President laughed alone. "Well," he said sobering and pulling his watch, "let it go. I think the Lord in His own good time and way will work this out all right. God knows what is best."

Lamon again asked for a promise that the President would not go out after dark while the marshal was in Richmond. Usher shook hands with his old friend and turned to leave.

"Well," said Lincoln, "I promise to do the best I can toward it. Good-by. God bless you, Hill."*

On the subject of dreams, the guard Crook later recalled his midnight patrols outside the President's bedroom. In the stillness, with only the squeak of floorboards to punctuate his pacing, Crook often heard Lincoln moan in his sleep. "I would stand there and listen," the guard said, "until a sort of panic stole over me. At last I would walk softly away, feeling as if I had been listening at a keyhole."

On the day before Mr. Lincoln's appointment with destiny, General and Mrs. Grant arrived in Washington. This was on Thursday, April 13. The hero of the war wanted to go up to Burlington, New Jersey, to see his two children and, with Lee out of the way, Grant felt that Sherman and Meade could handle Joe Johnston. He stopped off in Washington only because Stanton wanted him to advise how to cut army personnel and to cancel certain army contracts. The general figured that he could do this chore in a day—or two at most.

The Grants were consciously unostentatious. They did not like the theater or parades or public appearances, and did

---

* The President could not have made such a promise in good conscience because he had learned, long before, that, when Mrs. Lincoln wanted to go out, he risked a scene if he refused.

not care much for dining out. At Appomattox Courthouse, Ulysses S. Grant expressly ordered that there be no victory celebration by the Army of the Potomac. Now this morning, he arrived at the Willard Hotel so quietly that the management was flustered. At the desk, he stood short and stocky and dusty, gray beard a little bit stained with brown, and explained that he wanted a sitting room and a bedroom for overnight. If he needed the suite for an extra night, he would let the management know. With the Grants were Colonel Horace Porter, the general's aide, and two sergeants who carried luggage.

In the rooms, Mrs. Grant unpacked and the general said that he and Colonel Porter would walk around the corner to the War Department and do some work. When the two stepped out on Pennsylvania Avenue, Grant was recognized and, in a trice, was surrounded by a hero-worshipping crowd. The people cheered. Porter, dismayed, tried to clear a passage for his chief. He found that he was helpless. Metropolitan policemen rescued the two officers and persuaded them to accept a carriage and a cavalry escort for the three-block trip.

At the War Department, the general was given a desk and, after a round of handshaking and congratulations, began the work of cutting the expenses of a wartime army. He recommended that the draft be stopped at once; he marked down the numbers of certain divisions and brigades which could be mustered out of service without impairing the power of the army to enforce the peace; he labored over contracts for shovels and ambulances and ammunition and beds and blankets and bullets which, in his estimation, would not be needed.

In the afternoon, Grant received an invitation from Mrs. Lincoln to take a drive around the city in the evening with her and her husband. The general did not want to go. He knew little about the social amenities—barely enough to make him fretful about his rights in such matters—and he went into Stanton's office and told him about it. Stanton said that the general might refuse on the grounds of impending work. Grant followed this advice, although he might have wondered why his wife was not invited.

In the late afternoon, Stanton was leaving the War Department when he stopped in to say good night and to remind the general that he and his wife were expected at an informal at-home with the Stantons. Grant said that he wanted to finish a few more items on his list of recommendations, and that he and his wife would be at the Stanton home later. He told the Secretary of War that, while he had successfully turned down the invitation for an evening drive with the Lincolns, he now had a second one—this from the President. Lincoln wanted him to attend the theater tomorrow night.

Stanton was irritated. In the presence of telegrapher Bates, he urged Grant not to attend. He said that he and other Cabinet members had warned Lincoln about these public appearances many times, and that he, Stanton, had made it a rule to turn down all such invitations. Washington City, the Secretary of War said, was "Secesh" to the core, a place of wild-eyed plots and explosive Southern temperament. Stanton urged the general to refuse the invitation and asked him to use his good offices to keep Lincoln from attending the theater.

What happened at this hour—6 P.M.—is not altogether clear, but it is important. The invitation to take the evening drive probably arrived from Mrs. Lincoln shortly after lunchtime. After it was declined (the White House record shows that Colonel Horace Porter visited the President's office in mid-afternoon), it seems credible that Mrs. Lincoln pressed upon her husband the public adulation being accorded to the general, and asked him to invite Grant to the theater as a means of giving the people a chance to look at the hero of the hour. The President told several friends on this day and again on Friday that he had no inclination to go to the theater himself, but felt that the public was entitled to see the general.

Although there have been explanations of this matter, and Grant added to them years later, the weight of evidence would indicate that the general, courageous in battle, lacked the courage to decline the second invitation. At one point in the conversation with the Secretary of War, the general said that "it was embarrassing to accept" the invitation to the theater. The words "it was embarrassing to accept" sound as though he had already accepted.

Still, Grant did not want to go. And neither did Lincoln.

That night, the Grants were entertained by the Stantons and two soldiers stood guard outside the house opposite Franklin Square. Some strollers asked who was inside and the soldiers, with pride, said General Ulysses S. Grant. A crowd collected and set up a clamor for a speech. The Secretary of War came out, and uttered some appropriate words from the steps. The general waved to the crowd and said nothing.

The social broke up early because both men had work to do. Grant said that he wanted to finish his task tomorrow, and be off for Burlington. Stanton said that he wanted to continue work tonight on a paper that the President had asked him to draw up in time for tomorrow's Cabinet meeting.

The secretary worked very late that night. The President retired early.

The Conspiracy

It is likely that John Wilkes Booth first decided to dispose of Abraham Lincoln the day after the presidential election of 1864. The actor despised the President before that; in the campaign that year Booth predicted that, if Lincoln was elected, he would set up a dynasty. Lincoln had been Booth's emotional whipping boy for at least four years.

There is no record tracing the origins of Booth's opposition. Somewhere, it had a beginning. These men had never met. Their personal paths never crossed, and Booth sought nothing that could be termed a favor. Still, abiding hatreds start somewhere, and it may be that Abraham Lincoln offended the actor by proclaiming, in the election of 1860, his intention of holding the Union together against the wishes of the Southern secessionists. Booth, an adolescent in politics, pictured the South as a land of courtly and proud people; the North, to him, was a land of crude mercenaries of enormous brute strength.

Whether it was this, or an accumulation of presidential acts designed, by necessity, to bring the South to its knees, no one knows, but it was a passionate and violent hatred of the self-feeding type. Lincoln had to do no more than breathe to cause John Wilkes Booth to loathe him the more each day. Wilkes argued with his family about Lincoln, was stunned when he learned that his brother Edwin had voted for Lincoln in 1860 and would do so again in 1864.

There is substance to the story that Wilkes hated Lincoln so much in 1864 that he was certain that his feelings were shared by a majority of the electorate and he was sure that the President would be turned out of office. Booth's venom was so strong that he found it impossible to understand people who had a kind word for Lincoln. In this he was sincere.

He was not insane—if his acts and his conversation can be weighed psychologically—any more than another man might be called psychotic for fearing snakes or wasps to the point of becoming a nuisance on the subject. He was emotionally immature—his sexual excesses and his inability to take orders alone tend to give one that impression—but he was also shrewd and generous and a loyal friend.

Above all, Booth had pride. He thought of his family as one of the finest in Maryland although his father had been called an insane alcoholic, and had not married his mother until years after the first babies had been born. His love for his sister amounted to melancholy adoration. His affection for his older brother Edwin was, for political reasons, more restrained.

Wilkes's father, an emotional sentimentalist, had studied the Koran, occultism and Catholic theology and, in his spiritual confusion, believed that all animals were reincarnated humans, so that, when a sparrow fell at Bel Air, Mr. Booth, Sr., gave it a complete funeral service. In anger, young Wilkes once killed a litter of kittens and the mother cat. His father wept uncontrollably. At another time, in summer, Wilkes forced a horse to pull him to town and back in a sleigh to win a bet.

In 1855, at the age of seventeen, he made his debut at the St. Charles Theatre in Baltimore. He put more fire into the role of Richmond than the part required, but the audience, which remembered the lines better than he, hissed him. After that, he studied harder, but he would always be known more for his spirit and his acrobatics than for his measured cadence. Two years later, Asia Booth Clarke prevailed upon her husband to give the boy a chance at the Arch Street Theater in Philadelphia, and John Sleeper Clarke quickly found himself with one more actor. The boy developed into an outrageous scene thief, but he played his parts with such heightened enthusiasm that the audiences idolized him.

By 1860, he was an established theatrical star and toured the South and West to jammed houses, although he was only twenty-two. His backstage conquests were buzzed from New York to Philadelphia to Baltimore to Washington to Richmond to Columbus, Georgia. In Madison, Indiana, pretty Henrietta Irving slashed at the star with a knife, missed, and plunged the blade into her own breast. Wilkes took his women as he took his brandy, in long careless draughts, and tossed the empties on a refuse heap.

John Wilkes Booth always played at love and always carried small photos of his special girls. One was Bessie Hale, the plump, dark daughter of Democratic Senator John P. Hale of New Hampshire. Another was Ella Turner. She is the only girl, a tiny redhead, who enjoyed the semi-permanent status of mistress to Booth. When he wasn't at the National Hotel, he ordered her to stay at her sister's house of prostitution on Ohio Avenue, around the corner from the White House. He never carried Miss Turner's picture.

Booth was a Southerner by choice. Geographically, his roots were in northern Maryland, at Bel Air, and, although many Marylanders served the Confederacy in the War Between the States, the state itself remained officially Northern. Booth did not enlist to fight for the cause he loved. He had served for a little while in a Virginia company which took part in the capture of John Brown, and the records of that company indicate that Booth may have been present at the hanging of Brown, but he was not under arms when the Civil War began. When friends asked why he did not enlist, he said that he had made a promise to his mother. Booth's apologists say that he had an unnatural fear of having his face scarred, and that this kept him from fighting in the Confederate cause.

Whatever the reason, his inability to fight and die for a cause he so fervently espoused could, in time, have become sickening to him. When Lincoln was elected for a second term, Booth decided that it was time to contribute, in a big way, to the Confederacy. He would kidnap the President. He would kidnap him, and smuggle him through the lines into Richmond. This plan, if executed, would achieve several results, all desirable to Booth: (1) it would make a historic figure of the actor; (2) it would humble the man he hated; (3) it would force the North to exchange prisoners with the South, something the man-rich North refused to do for the man-poor South.

The word "kidnap" is not Booth's. In his later dealings, he used the word "capture," and there is a difference because, in Booth's idealistic righteousness, kidnapping was a crime, capture in wartime was not. Lincoln, as commander in chief of the armies, was a soldier and a soldier is liable to capture.

Mr. Booth decided in November 1864, to work his scheme. He had no fear for his life. In his vanity, it was more important to be remembered forever than to live long. His paramount fear was that the world might misunderstand his great act; might attribute the deed to mean motives, and so he resolved to leave a note to the world, so that the history books of tomorrow and yet tomorrow would have the story straight.

In the late autumn, Booth explored the exits from Washington. There were four ways out of the city to the Southern states. The farthest from the White House was the Georgetown Aqueduct, which lay to the northwest. He declined this escape route at once because it entailed a ride of a mile and a half through the city to reach it and, once across the Aqueduct, he would be on Virginia soil and, being northwest of Washington City, he could be cut off easily by any patrol leaving the city in a southerly direction.

A second escape route, considered and discarded at once, was at the opposite end of town. This was Benning's Bridge, to the east. It lay beyond the poorhouse and would carry him across the east branch of the Potomac River (now called Anacostia River) into Maryland, but he would then find himself on the road to Annapolis, in a southeasterly heading rather than south.

The third possibility was Long Bridge, a few blocks south of the White House. This route aimed directly at Richmond, a hundred miles away. The virtue of this road was also its vice. The Army of the Potomac had been using it, on and off, for four years, and parts of it were heavy with the traffic of regiments and brigades heading for the front, or heading home for leave. Big sections of this country had been broken into Union area commands, and the War Department could get in touch with any of these by telegraph in a matter of minutes. Booth rejected Long Bridge.

There was but one other way. This lay across the Navy Yard Bridge, at the foot of Eleventh Street. On the far side was southern Maryland, a big peninsula of little villages and secessionist intrigue. Southern Maryland had been almost isolated from the war. No armies stirred the mustard dust of its roads; no villages were fired in raids; it was neglected by both of the big protagonists, except as a courier route for spies. Mr. Booth explored this route, from the Navy Yard Bridge down to Silesia and Pomfret and Port Tobacco, where, for a price, one could secure a boat and cross into Virginia slightly below Fredericksburg. He rode back up through Indian Head Junction and Bryantown and Surrattsville, stopping here and there to question the people and to pose as a farm buyer and a horse trader.

Booth decided that this was to be the escape route. He tested the sympathies of the people he met with adroit questions, and he came to feel that this was friendly country indeed. On one of his trips, he was between Waldorf and Bryantown and he was introduced to Dr. Samuel Mudd, a humorless farmer who had not practiced medicine in years. The Mudds, and their kin, owned a lot of good property in the area and they were anxious to make a sale to the actor. On one occasion, Booth was invited to spend the night at the home of Dr. Mudd. Still, Booth did not buy a farm and probably never intended to buy one. He did buy two horses from a Mudd friend—a small saddle horse and a large one-eyed roan with fetlocks like a brewery horse.

The first outward intimation of the type of crime Booth planned came in the winter of 1864–1865 when, on a trip to New York, he stopped on an icy evening to visit an actor friend, Mr. Samuel Knapp Chester. This man was a fair character actor who depended, for work, on friendly stars like Booth. He lived at 45 Grove Street with his wife and children.

On this evening, Booth stopped by and asked Chester to take a walk with him. They went crosstown, heads bent into the collars of their capes, and they stopped at the House of Lords, a tavern on Houston Street. They talked about the theater season, gossiped about fellow actors, talked about "side" investments in other businesses, and the chance of Chester getting some work at Ford's Theatre in Washington City, or at Grover's.

They ate and drank. Chester listened. He hoped that Booth would talk about his land speculations, because Chester had received a note from him telling about the big money to be made in southern Maryland farms and livestock. Mr. Chester never made a great deal of money and, between engagements, he was always hard pressed for cash. But Booth kept talking theater and, at last, Chester said:

"Tell me about your speculations."

"I have a new speculation," said Booth.

"I want to hear it."

"You will—by and by."

There was no point in pressing Booth. He would talk about it, Chester knew, when he was ready. They left the House of Lords and stopped at an oyster bar under the Revere House and Booth was suddenly quiet. They ate and drank considerably. Booth was weighing a momentous decision—whether to tell this man or not. Once told, if Chester refused to participate in the conspiracy, Booth would be at the mercy of his friend.

The two men went out into the night and walked up Broadway, almost in silence. When they reached Bleecker Street, Chester said that, if Booth did not mind, he would turn west here and go home. Booth asked him to walk a little farther. At Fourth Street and Broadway, Booth said: "Broadway is still crowded. Let us walk down Fourth."

After a block or two, with no one in sight, Booth stopped under a streetlamp. He looked at his friend and began a preamble about how few friends a man could really trust in these days, and, when Chester saw that his patron seemed to lack the confidence to continue, he said: "For God's sake, Wilkes, speak up!"

Booth blurted out that he was now engaged in a conspiracy to capture the heads of the United States Government, including Lincoln, and he planned to bring them to Richmond. There was a silence. Chester, fumbling for words, afraid to believe the ones he had heard, said: "You wish me to go in this?"

"Yes," said Booth.

"It is impossible, Wilkes." Chester held out his hands in supplication. "Only think of my family."

"I have two or three thousand dollars I can leave for them."

"No," said Chester. He was shocked and bewildered. He had expected a business proposition, a chance to make some "side" money, and now he was being offered a chance to help perpetrate a deed so foul that he had to keep staring at Booth to be assured of the seriousness of his friend. It was beyond comprehension. He wanted to go home.

Booth wanted to talk. They stood under the lamp for twenty minutes. The sum of Booth's argument was that the North had forced the South into war, that the South wanted, in honor, to exchange prisoners of war with the North but that Lincoln had refused to do it. The reason for this, Booth explained, was that the North had an inexhaustible supply of men while the South was now using fourteen-year-old boys, in uniform, to guard prisoners. The South did not want prisoners and could not afford to feed them and, in the light of the embargo, could not even give them medicine. All the South wanted was a fair exchange—man for man. Each day, she grew a little weaker. Each day, she faced defeat unless, in some way, the President could be forced to agree to an exchange of prisoners.

Booth said that he could assure Chester that there would be a fortune, and honor, for those who helped to capture Lincoln. A man could find himself rich almost overnight.

"No," said Chester.

"Then you will not betray me?" said Booth quietly. "You dare not."

"You do not have to be afraid of me, Wilkes."

"I will implicate you anyhow."

"That is unnecessary."

"Our party is sworn to secrecy, Sam, and if you betray us, you will be hunted down through life."

"I will forget all that you have said."

"I urge you to come in with us."

"No."

"Your work will be simple, Sam. You understand theaters. All you will have to do is open the back door of Ford's Theatre at a signal."

"In Washington?"

"It is easy and you will succeed."

"Wilkes, please. I have a family."

"Your family will get good care. We have parties on the other side who will co-operate with us. There are between fifty and a hundred people in this."

"Wilkes, I must say good night." Chester went home.

The shreds of evidence, held together, say that John Wilkes Booth was lying. At best, there were never more than seven persons in his plot. In the main, they were simpleminded schemers, not one of whom rose above the rank of private in the Confederate Army. Each—with one exception—had a greater personal loyalty to Booth than to the South. None had qualities of leadership. So far as parties on "the other side" are concerned, there were none. Booth wanted none. He wanted to do this thing alone, with the assistance of courageous men smaller in stature than he. At no time did he seek official sanction, or even unofficial sanction, from the South.

Booth was a loner.

Between January 1865, and April, the conspirator put about $4,000 of his money into the "capture." The biggest part of this was spent for supporting his fellow conspirators; a little went for horses and feed, and some went for fleeing Washington when successive plots failed.

These plots were movements of opportunity. In retrospect, some of them have comic aspects. Throughout January, February and March, the element of coincidence was on the President's side and, as each plot failed, the conspirators felt that the failure indicated that the government was aware of the Booth band; this bred panic, and the group dispersed. At no time, with one exception, did the United States Government know about the conspiracy and, on that occasion, the administration gave it little attention.

The first attempt at "capture" was scheduled for the night of Wednesday, January 18. It had been announced that Mr. Lincoln and two friends would attend Ford's Theatre to see Edwin Forrest in *Jack Cade,* a play about the Kentish revolution.

A few days before, Booth stopped at the home of his sister Asia, in Philadelphia, to sign an important letter he had left in her keeping. It was to be released only if he was captured or killed. This was the letter to the history books. It is a rambling document, replete with the customary calls to God to bear witness, the breast beating, the indictment, heartbreak, mother and flag. The substance of it is as follows:

Right or wrong, God judge me, not man. For be my motive good or bad, of one thing I am sure, the lasting condemnation of the North. I love peace more than life. Have loved the Union beyond expression. For four years have I waited, hoped and prayed for the dark clouds to break and for a restoration of our former sunshine. All hope for peace is dead. My prayers have proved as idle as my hopes. God's will be done. I go to see and share the bitter end.

I have ever held the South were right. The very nomination of Abraham Lincoln, four years ago, plainly spoke war, war upon Southern rights and institutions. His election proved it. "Await an overt act." Yes, till you are bound and plundered. What folly. The South was wise. Who thinks of argument or pastime when the finger of his enemy presses the trigger? In a foreign war, I too could say "Country right or wrong." But in a struggle such as ours (where the brother tries to pierce the brother's heart) for God's sake *choose* the right! When a country like this spurns justice from her side, she forfeits the allegiance of every honest

freeman, and should leave him, untrammeled by
any fealty soever, to act as his conscience may ap-
prove. . . .

The country was formed for the white, not the
black man. And looking upon African slavery from
the same standpoint held by the noble framers of our
constitution, I, for one, have ever considered it one
of the greatest blessings (both for themselves and
us) that God ever bestowed upon a favored nation.
Witness heretofore our wealth and our power; witness
their elevation and enlightenment above their race
elsewhere. I have lived among it most of my life, and
have seen less harsh treatment from master to man
than I have beheld in the north from Father to son.
Yet, Heaven knows, no one would be willing to do
more for the Negro race than I, could I but see the
way to still better their condition.

But Lincoln's policy is only preparing a way for
their total annihilation. The south are not, nor have
they been, fighting for the continuation of slavery.
The first battle of Bull Run did away with that idea.
Their cause since the war have been as noble and
greater far than those that urged their fathers on.
Even should we allow that they were wrong at the
beginning of this contest, *cruelty* and *injustice* have
made the wrong become the right, and they stand
now (before the wonder and admiration of the world)
as a noble band of patriotic heroes. Hereafter, reading
of their deeds, Thermopolae will be forgotten. . . .

The south can make no choice. It is either exter-
mination or slavery for *themselves* (worse than death)
to draw from. I know *my* choice. . . .

But there is no time for words. I write in haste. I
know how foolish I shall be deemed for undertaking

such a step as this, where, on one side, I have many friends and everything to make me happy, where my profession *alone* has gained me an income of more than twenty thousand dollars a year, and where my great personal ambition in my profession has such a great field for labor. On the other hand, the south have never bestowed upon me one kind word; a place now where I have no friends, except beneath the sod; a place where I must either become a private soldier or a beggar. To give up all of the former for the latter, besides my mother and my sisters, whom I love so dearly (although they so widely differ with me in opinion) seems insane; but God is my judge. I love justice more than I do a country that disowns it, more than fame or wealth; more (Heaven pardon me if wrong) more than a happy home. . . .

My love (as things stand today) is for the south alone. Nor do I deem it a dishonor in attempting to make for her a prisoner of this man to whom she owes so much of misery. If success attends me, I go penniless to her side. They say that she has found that "last ditch" which the North has so long derided, and been endeavoring to force her in, forgetting they are our brothers, and that it is impolitic to goad on an enemy to madness. Should I reach her in safety and find it true, I will proudly beg permission to triumph or die in that same "ditch" by her side.

A confederate doing duty upon his own responsibility.

J. Wilkes Booth.

He signed it, assured himself of a place in the history books, and hurried back to Washington. He sent to Baltimore for two boyhood friends, Michael O'Laughlin and Sam Arnold. Both were Confederate veterans, and both were hardened to the rigors of war, but they were shocked when their old friend told them the mission. Arnold was so frightened that he spent time trying to convince Booth that the scheme had to fail.

Neither of these recruits was bright. They were poor Baltimore boys who looked upon Wilkes as a rich and influential friend. The actor convinced them that they were part of a big secret band.

The rest of the group—Arnold and O'Laughlin had not met them yet, nor even heard their names—consisted of George A. Atzerodt, a carriage maker from Port Tobacco; David Herold, a young drug clerk who wearied of a matriarchal world; and John Surratt, Confederate courier, whose mother managed a boardinghouse.

Mr. Atzerodt is worth some special comment here, since he was later "assigned" to kill the Vice President. He was a German who worked by day with wood and wheels—a small man with small sly eyes and a drooping mustache; a man with features as malleable as warm putty; a man who always looked dirty and was conscious of it. At night, he ferried Southerners back and forth across Pope's Creek and, if a Northerner wanted to get through the blockade, George would ferry him too. The kindest thing that was ever said about Mr. Atzerodt was that he was a man who would not resent an insult.

He was pitifully anxious to make a friend, and to this end he bought drinks for barflies and laughed at their jokes, but, the moment any of them challenged something that he had said, Mr. Atzerodt jammed his brown beaver hat on his head and left.

Booth had five men, in two groups. Each was in the plot to "capture" the commander in chief of the Union; none wanted to kill; two were in serious doubt about the propriety and feasibility of capture. Sam Arnold was afraid of any plot involving

Lincoln. John Surratt, who had risked his life for the Confederacy as a courier, started as a member of the band by entertaining the notion that the arch-conspirator was insane. However, Booth visited the H Street boardinghouse and charmed Surratt with his candor and absence of condescension, and convinced the courier that the very brazenness of the idea would help to effect complete surprise, plus the fact that "capture" was a legal act.

Still Booth needed an actor, a theater-wise person who could turn out all the lights in a theater on cue, and, having been turned down by Sam Chester, he tried to enlist the services of a small-parts actor in Washington named John Matthews. Mr. Matthews was conscious of his own smallness in the world of the American theater, and, although he worked the full season at Ford's Theatre, his habit was not to drink with actors at Taltavul's saloon because he might be expected to buy drinks in return. He drank in a small place a block away from the theater.

Booth tried to interest Matthews in the plot and the little actor recoiled. He turned it down at once and advised the star to forget it. "Matthews," said Booth later, "is a coward and not fit to live." The actor would not forget his contempt of Matthews, and would try to hurt him.

The conspirators—with the exception of Arnold and O'Laughlin—met infrequently at Mrs. Surratt's boardinghouse. They whispered, consulted in upstairs rooms, wrestled with knives on a bed, bought pistols and became acquainted with their workings, and rode off into the country. The widow Surratt got to know them and once, in a moment of reflection, she asked her son John why these men were trooping into the house at odd hours and John said that they were all interested in a common oil speculation. Mrs. Surratt admired Booth, the courtly gentleman who attracted the eye of her seventeen-year-old daughter; she was fond of young David Herold, who was full of tall tales of hunting in her own southern Maryland; she didn't like George Atzerodt, whom the boarders called "Port Tobacco." Mrs. Surratt was, by all the rules of evidence, a pious zero with a penchant for falling on evil days. There is no corroborative evidence to show that she ever knew anything about a plot.

Mrs. Surratt had three children: Isaac, a Confederate soldier; John, a Confederate courier; and Anna. The boardinghouse kept Mrs. Surratt and Anna alive. Years before, she and her husband had had a farm and a tavern in southern Maryland and the government had made Mr. Surratt a postmaster and had called the crossroads Surrattsville. A few years ago, Mr. Surratt had died and his widow learned that the farm and tavern were difficult to administer. She had called John home from St. Charles College, near Ellicott's Mills and, for a while, the boy filled his father's shoes as a local postmaster. The appointment went to someone else, and John found that the rest of it had no appeal for him. He was a tall, blond, intelligent boy with cavernous eyes and a domed forehead. He was now twenty, and so he grew a wispy goatee.

Mrs. Surratt leased farm and tavern to Mr. John Lloyd, a drunkard with a poor memory. She took John and Anna off to Washington City and opened her impeccable little boardinghouse on H Street. She placed advertisements in the *Star* and the *National Intelligencer* and she got boarders and set a good table.

Still, her troubles were economic and she needed every penny due her. She was in debt, for example, to Mr. Charles Calvert of southern Maryland for a few hundred dollars. In protracted correspondence with him, she held him off by saying that, many years ago, her husband had sold a piece of property to Mr. John Nothey and, if she could get him to pay her, she would be happy, in turn, to pay Mr. Calvert.

Her political horizon was small, and it is doubtful that she understood the issues between the states, but it is beyond argument that her sympathy was with the South and she was certain that the North was wrong in invading the South. She had owned a few slaves at one time, and at least one of them testified that she was harsh; two others testified that she was warm and solicitous. It is known that, at Surrattsville, she had fed passing Union soldiers and refused to accept money for it. Once she found some stray army horses and she had barned them until the proper authorities called for them. She refused to accept payment for feeding them.

Among her boarders now, all of whom had eaten early today, were Mr. and Mrs. John T. Holahan, and their daughter, fourteen. Mr. Holahan was a big man with big hands. His work was the cutting of tombstones. The Holahans occupied the front room on the third floor and the alcove too. The back room on that floor was used by John Surratt (when he was at home) and a former schoolmate, Louis J. Wiechman. Mr. Wiechman was big and soft and pungent, an overripe melon. He had studied for the priesthood at St. Charles and had failed. He had taught in school for a while, but that job too had sifted through his hands. Now he worked for the United States Government at the Office of the Commissary General of Prisoners. Some of his failures may have been attributable to Wiechman's personality, which was akin to that of a professional sneak. He felt drawn to eavesdropping and gossip and, at the same time, had the aura of a suffering saint who has been snubbed.

There was a low-ceilinged attic in the boardinghouse and this was used as a bedroom and dressing room by Miss Anna Surratt and her cousin, Olivia Jenkins. Both were young and coquettish and bought postcard photos of actors and brave Southern generals.

On the second floor there was a sizable sitting room—

which was reached from the outside of the brick house by a white inverted V staircase—and a back parlor. This parlor was used as a double bedroom by Mrs. Surratt and a young boarder, Miss Honora Fitzpatrick. On the ground floor—or basement—was another sitting room, a dining room, and a kitchen.

Little evidence remains of this first attempt to kidnap President Lincoln. On the weekend prior to the Wednesday of the attempt (January 18) Herold was sent to southern Maryland to arrange for relays of horses. Atzerodt was in Port Tobacco inquiring about leasing a flatboat large enough "to float ten or twelve people and a carriage."

The mechanics of the kidnapping appear to have been that Surratt would be detailed to shut off the master gas valve, under the stage of Ford's Theatre, at a signal. This would extinguish every light in the theater. He was then to come up onstage in the dark and wait, as Booth, in Boxes 7 and 8, forced the President at gunpoint to submit to gag and ropes. The actor would lower the President over the façade of the box eleven feet to the stage, then lower himself to the stage. The two men would hustle the President offstage, out the rear door, where a covered wagon would be waiting in the alley. There would be some confusion in the dark theater, among actors as well as patrons, and Booth counted on this to assist, not to hinder him. The President would be placed in the back of the wagon, trussed, and Surratt would drive the wagon out of the alley with Booth riding single-mount behind the wagon.

On the far side of the Navy Yard Bridge, they would pick up the first of Herold's team relays, and head for Port Tobacco, twenty-nine miles away. By the time Atzerodt had ferried the party across the Potomac to Mathias Point, the whole country would know of their glorious deed and the people of Virginia would assist them through the battle lines to Richmond.

Arnold and O'Laughlin were not part of this attempt. As punishment for not showing sufficient enthusiasm, Booth proceeded without them. At 7 P.M. on January 18, the plotters were ready.

President Lincoln did not attend the theater that night. No reason was given. The management of the theater expected him because the partition between Boxes 7 and 8 was taken down in the afternoon and the President's favorite rocker was placed in the part of the box closest to the dress circle.

The disappointment was almost too much for Booth and his little band to bear. The following morning, they scattered like minnows. Booth fled to New York. Surratt went south to the protection of the Confederacy. Herold hurried back to his mother and his seven sisters. Atzerodt took a job in Port Tobacco.

In early February, the band took slight heart. There had been no arrests, no apparent shadowing. John Wilkes Booth enlisted the final member of the conspirators. In a way, this man was the best because he could be relied upon to kill on order. His name was Lewis Powell and he was a native of Florida. He had changed his name to Lewis Paine, and he would be known by this name until he died.

Lewis Paine was big and strong and silent and stupid. He had thick jet hair, a clean, handsome face, and the muscles of a circus strong man.

In the South, he had seen John Wilkes Booth on the stage once. Afterward, he had been taken backstage to meet the star, and Lewis Paine never forgot the courtly manners, the gracious attitudes, the born-to-rule air. Later, Lewis went off to war with his brothers and he developed into a most efficient soldier. His quiet boast was that he had never wounded a Union soldier. He killed—or missed. His greatest shield against the moral strains of war was his stupidity, which kept him doing the work he was ordered to do, while preventing him from pondering on it. With no boastfulness, he displayed a skull which he used as an ash receiver and said that it was the head of a Union soldier whom he had killed.

Paine fought hard and well in the Peninsula Campaign, at Antietam, Chancellorsville, had two brothers killed at Murfreesboro, fought again at Gettysburg, was wounded and taken prisoner.

At this point, there is an unexplained hitch in his record. Paine was assigned as a male nurse in a Union hospital and escaped. He was next seen in the city of Baltimore, where Union authorities, instead of arresting him, ordered him to move farther northward, to Philadelphia or New York. It may be that it was here that Powell changed his name to Paine, and the authorities, having no record of Paine, assumed that he was one of the many wandering deserters of the Confederate Army, and wanted him at least two hundred miles north of the battle lines.

He was twenty, and boarded with Mrs. Mary Branson at 16 Eutaw Street, Baltimore. A few of the neighbors tried to make friends with him, or to strike up an acquaintanceship, and these drew nothing more than a blank stare. When he talked, he seemed to do it without moving his lips. Only the right side of his upper lip showed motion, and this gave him a sneering manner.

Paine's weakness was a rare temper. It seldom mastered him, but, when it did, mastery was complete. A Negro maid came into his room one morning to make up his bed and he asked a question. She made the mistake of answering insolently. In a flash, both of his big hands were around her throat and he squeezed until she collapsed and fell to the floor. He stood over her, staring. The maid lived.

In the second week of February, John Wilkes Booth was in Baltimore to see Arnold and O'Laughlin about resurrecting the "capture" when Paine, lounging on a street corner, saw him and hailed. In spite of almost five years of time, Booth remembered the big Southern kid who had once been brought backstage for an introduction. They had a long talk about Paine's war record, and the actor bought him a suit of clothes and gave him money.

From that moment on, Booth had a faithful dog. Paine's feeling for the actor was slightly shy of idolatry. Booth was pleasantly surprised to find that his new man was almost ideal; he would do as he was told without question; he could be left alone for weeks in a boardinghouse and would not get into mischief and did not care for the company of girls. He seemed to be able to spend long periods of time sleeping and eating.

Paine was brought to the Surratt boardinghouse and introduced as the Reverend Lewis Wood, Baptist preacher. Mrs. Surratt, Catholic, thought that it was amusing that a Protestant minister would seek her place, of all the boardinghouses in Washington, but she told her daughter Anna that if the Reverend had no complaints, she had none.

There was much to see in Washington, but Paine was not interested. Every time he ventured on the streets to reach a rendezvous with "Cap," he got lost and found it difficult to get back to the boardinghouse. He complained that the streets were laid out crooked, that they did not intersect at right angles, and he could make no sense of them.

On the night of Tuesday, March 7, four weeks and three days prior to the important day, a small incident occurred at Ford's Theatre. Mr. Thomas Raybold, ticket seller, sold four orchestra seats in advance to Thomas Merrick, the day clerk at the National Hotel. The policy of the theater, when seat holders did not show up by the end of the first act, was to permit the ushers to move less favored patrons up to the empty chairs.

Merrick arrived, at the start of Act Two, to find that his seats had been taken. With him were a Mrs. Bunker and a Mr.

Norton. Merrick was irritated. Raybold too was distressed, and offered to show the party to any good seats in the house, box seats. This mollified the party and they followed the ticket seller up the dress circle stairs of Ford's Theatre and down the left-hand aisle to Box Number 6. It was locked.

Raybold's embarrassment deepened, and he explained that the usher kept the keys to all boxes, because Mr. Ford did not like to have the stagehands sleeping in them by day, but that the dress circle usher was home ill and the best thing to do would be to take the party to the other side of the theater and put them in the presidential box. He led the party to the back of the dress circle, across, and down the right-hand aisle. They went through a little white door to Box 7. The door was locked. Raybold tried the door to Box 8. Locked.

The ticket seller, at this point, was angry at himself. He placed his shoulder against the door of the box and pushed. The door bent inward, and bounded back. He pushed again. In the rear of the dark corridor, Mrs. Bunker giggled. Raybold lifted his foot, aimed at the lock, and smashed. The lock snapped. The door flew open. The hasp which had held the lock swung loosely. When the party had been seated, and had forgiven him, Raybold tried the lock and found that it was broken. In the future, the door to the presidential box could be opened by anyone.

Mr. Raybold did not report it.

On a cold afternoon, Booth took Lewis Paine for a walk and showed him the White House. They walked across the south grounds toward the front of the mansion. Booth talked confidentially as they looked at the black stately trees, the squatters, the sentries warm inside their boxes, and heard the complaining bleat of Tad Lincoln's goats, the barking challenge of a sentry on the far side of the mansion.

"He is right over there," Booth said, pointing. Paine looked. "If you really want to kill him," the actor said, "what I would do is just walk in, present my card and, when I was admitted, walk up to his desk and shoot him."

Paine made no answer. Booth said he lacked nerve. Still no answer. The actor offered a less dangerous alternative. If you want to, said Booth, you can lie in wait in the bushes at the front of the White House lawn any evening and shoot him as he returns from his last daily visit to the War Department.

Lewis Paine liked that idea. He said he would do it. This is the first time on record that the thoughts of John Wilkes Booth turned from capture to kill.

The soldier waited in the bushes one night and, when he returned to the boardinghouse, he told Booth that he had lost his nerve. He insisted that he had been close enough to have strangled the President of the United States.

Lincoln had walked back to the White House that night with Major Thomas Eckert, the chief of telegraphers, and Paine had heard the President say, in a jocular way: "Major, spread out, spread out or we shall break through the ice."

Sometime in March 1865, the clerks in the office of the Commissary General of Prisoners were talking about the illness of the President, and some fell to wondering what would happen to the Union if he died. This, in turn, led to a discussion of the assassination plots featured in the newspapers and Louis Wiechman, the fat boarder at Surratt House, assumed the air of a man who has a rich morsel of gossip and said that a plot was hatching against Mr. Lincoln in the very house where he boarded. A group of "Secesh" people were scheming to do away with the President.

Unless Wiechman's character is being read wrong, this was intended as thrilling gossip, nothing more. Had the boarder feared for the President's life, he might have been expected to report directly to his superior at the Commissary of Prisoners, or, conceding Wiechman's flair for the dramatic, he might have gone directly to Secretary of War Stanton. The least he might have done was to report the matter to Major General Christopher C. Augur, Commander, Department of Washington, 22nd Corps.

Wiechman was surprised and worried when he found that his morsel got out of hand. He was questioned by Captain Gleason of the office, who said that he would report the matter at once to Assistant Provost Marshal Lieutenant Sharp. Wiechman was worried. He was intelligent and he may have feared that someone might suspect that he was part of the plot. He hurried at once to a nearby office and breathlessly reported the entire matter to Captain McDavitt, U.S. Enrolling Officer. Thus Wiechman was on record as having patriotically warned the nation of the impending peril, even though he later admitted that he "talked secesh, but it was buncombe," and even though it was proved that, after exposing the plot, he entertained Atzerodt in his bedroom and lent his military coat and cape to Atzerodt and Paine, and he continued to share a bed with John Surratt.

Louis Wiechman told Captain McDavitt the names of all the habitués of the boardinghouse, as well as the residents, so the government was armed with information. No captain would, on his own authority, withhold such information. It can be assumed that it boiled upward toward Stanton. Captain Gleason had the same information, independently, and he brought it to the attention of Lieutenant Sharp, who also sent it to higher echelons for evaluation. At the top of both heaps was Stanton, who was so chronically worried about assassination attempts that he was seeing plots where there weren't any. Is it too much to suggest that the United States Government, on one level or another, was aware of John Wilkes Booth and his band, plus the boardinghouse at 541 H Street, in mid-March of 1865? Is it too much to expect that the government officers would give this report more than casual attention because it could not be classified with the crackpot anonymous letters which usually told about such plots, but came, rather, from a trusted clerk who worked for the War Department?

Mr. Stanton had caused the arrest of 38,000 persons in the war years, many on far flimsier evidence than the word of an army informer. Besides, Stanton was almost always in an arresting mood and, with the suspension of the writ of habeas corpus, it would have required only a nod to put Booth and his band, and the Surratts too and their boarders, behind bars. In separate cells, under interrogation, no one can doubt that Atzerodt—and perhaps Herold too—would have cracked and told the story of the "capture" within a day or two.

Nothing was done, although detectives would insist later that they had the Surratt boardinghouse under surveillance for weeks. The safest surmise is that both reports, Gleason's and McDavitt's, were read, assessed and filed somewhere on the road up. They were never found. The detectives who said that the house had been watched for weeks were asked to relate the daily habits of any of the boarders, and couldn't.

To Booth, time was running out for the Confederacy. Whatever was going to be done would have to be done quickly. All the war news was, to him, tragic. The South was collapsing and, if he didn't hurry, the war would be over and the Confederacy would be dead and there would be no cause to help.

On Monday, March 13, he began to call his band together and he started by sending a telegram to Mike O'Laughlin:

MR. O'LAUGHLIN
57 NORTH EXETER STREET
BALTIMORE, MD.
DON'T FEAR TO NEGLECT YOUR BUSINESS. YOU HAD
BETTER COME AT ONCE.

J. BOOTH

By Friday, Booth was ready. For the first and only time, all of the conspirators were together. He asked both groups to meet him at Gautier's Restaurant, 4 1/2 Street and Pennsylvania Avenue, at midnight. Early that evening, he asked John Surratt to take Paine to Ford's Theatre so that the ex-soldier could become acquainted with the premises. Surratt rented a closed carriage and took Paine, in Wiechman's military cape, and two ladies, Miss Honora Fitzpatrick, nineteen, and Miss Appolonia Dean, eleven years of age. All sat in the President's Box. Booth had determined that Paine was the man who had the nerve to kill and that, if there were others to kill or capture besides Lincoln, he wanted Paine with him.

At intermission, Booth appeared in the dark doorway and asked if everybody was enjoying the play. He called Paine and Surratt out into the dark corridor behind the boxes. In ten minutes, they returned. When the final curtain dropped, Surratt and Paine took the ladies back to the boardinghouse, and then proceeded to Gautier's Restaurant for the meeting.

Booth had engaged a private dining room. Cold cuts and cheeses and bottles of whiskey and champagne had been set up. The actor had asked Mr. Lichau, who owned Gautier's as well as Lichau House, to please see that this party was not disturbed. At it, everybody except Herold drank freely. Booth drank champagne and did most of his talking standing up.

He introduced each conspirator aloud, pointing his finger at each man in turn. The Arnold-O'Laughlin wing had not met the Herold-Atzerodt-Surratt-Paine group until now. Except for Booth, no one seemed to have any heart for the "capture" of Lincoln. Most of them, when permitted to speak, said that the government was already aware of the plot and that it would be dangerous to proceed.

Atzerodt sat with his brown beaver hat on, chewing on a cigar, his tiny eyes darting from face to face. Surratt, slender and pale, studied his drink and listened. Paine, big and blank, tried feebly to mask his contempt of the others. Sam Arnold, opposed to the whole thing and frightened as well, was obliged to Booth for the suit he was wearing. Mike O'Laughlin, in checked trousers and fawn-colored longcoat, stroked his long mustaches and drank his whiskey neat; Herold, happy to be a man among men, paid rapt attention to his idol.

Booth talked on. He admitted the difficulty of removing the Chief Magistrate, against his will, from Ford's Theatre. Paine, he said, would assist him in the State Box. The two of them would truss the President, screened from the audience by the folding drapes on the box and, at a signal, Sam Arnold would walk onstage, with drawn gun, and wait below the box for them to lower the burden. At this time, Mike, below stage, would shut the petcock and the gaslights would go out. Surratt and Atzerodt would be waiting on the far side of the Navy Yard Bridge and would lead the whole party to the flatboat. Davey Herold would sit on the driver's seat of the covered carriage behind Ford's Theatre and would drive out of the alley as Arnold jumped in back with Lincoln. O'Laughlin, Paine and Booth, on single mounts, would remain in the alley a moment with drawn guns, to hold off pursuit, and then would rejoin the wagon.

It was simple. Or was it? Booth finished speaking and fell silent. He waited for comment. None came. His skin began to whiten. At last, Sam Arnold coughed and opened his mouth. He, for one, he said, was opposed to the plan. He intended no offense to Wilkes, the most loyal of friends, but, in the first place, no one was ever sure when Lincoln would attend the theater. In the second place, there was no guarantee that the President would submit meekly to capture. Third, it would be dangerous to work under the noses of a thousand witnesses. Fourth, the entire North would be alarmed immediately after the deed and the group would be captured within an hour.

He didn't mention the fifth item: that it was apparent to

everyone that the South had lost the war and that the capture of Lincoln would be dramatic and pointless. Arnold looked up and saw that Booth was standing almost over him. Without looking alarmed, Sam fingered his drink and said, mildly, that it would be better for all concerned if the capture could be arranged to take place in the suburbs. He had read somewhere, he said, that the President was scheduled to attend a matinee at the Soldiers' Home, away out on Seventh Street, and it would be much easier to stalk a carriage on a lonely road, beat the guards into submission, and run off with Lincoln. In that way, he added, it would be quite some time before the government became aware of what had happened, and so the band would be well on their way into southern Maryland before an alarm could be sounded.

Booth was pale.

"Another thing," Arnold said. "If this thing isn't finished within a week, I am going to withdraw."

"Any man," said Booth slowly, "who talks of backing out ought to be shot."

All eyes turned on Sam. He shrugged, looked at his drink, and then smiled up at his patron.

"Two," he said, "can play that game."

Booth stood shaking, subsided, drank another glass of champagne, and apologized to Arnold. Free discussion was encouraged, and most of the men favored Arnold's plan. At dawn, Booth surrendered to his men. All right, he said in effect, if Lincoln is going out to Soldiers' Home it will be a matinee performance and I will hear about it at the theater, because some of our players help out at Soldiers' Home. I am appearing at Ford's Theatre tonight (Saturday, March 18) and I will get whatever news there may be. I will pass the word after the performance.

That night Booth played the part of Pescara in *The Apostate*. He learned that Sam was right. Some of the Ford troupe had been booked to play a matinee at Soldiers' Home on Monday.

It was a busy weekend. Surratt and Atzerodt rode to Surratts-ville, where they met Herold. The three drove southward five miles to a village called T.B. An hour later, they returned and sat in the tavern playing cards.

Surratt took John M. Lloyd, the alcoholic who had leased the tavern from Mrs. Surratt, to the front room. There, on a sofa, was a bundle of material. There were two army carbines with covers, a coiled length of hemp, a monkey wrench. Surratt asked Lloyd to hide the stuff for him. The tavernkeeper said that he wanted nothing to do with guns. Union patrols, he said, had been searching homes in the area looking for weapons and contraband, and he was not going to be found with guns.

John Surratt said that, when he had lived in the tavern, he had found an excellent hiding place over the kitchen, a tiny room with bare studs and beams. Lloyd, who had been all over the premises many times, doubted the existence of such a place. Surratt took the guns and rope, and led the man to the place. There, between the joists, the material was hidden.

"We'll pick it up in a few days," Surratt said.

On Monday morning, March 20, the final plans were laid. The conspirators left Washington City, on horses, in pairs. Arnold and O'Laughlin left first, at noon; Atzerodt and Paine departed next; Booth and Surratt last. Herold was stationed at the tavern in Surrattsville, waiting with the "stuff." He was told to get some axes too. In the event of close pursuit, the rear guard of the conspirators would fell trees and, after sundown, stretch rope across the road at low level. A boat was waiting at Port Tobacco.

Booth's only worry was a shortage of horses. None of the men had their own, although Surratt boasted to friends that he kept his own mounts at Howard's Stable on G Street. Booth owned two horses, the small trotter, and the big draft-type blind in one eye. Both of these were kept in a small stable behind Ford's Theatre.

There was a small, sagging building about a hundred feet south of the theater in the alley, and Ned Spangler, a stage-hand who had once worked for Booth's father at Bel Air, had renailed old clapboards on this building and had hung a new door. Spangler, an untidy man of brown hair and squinting eyes, was another in a parade of heavy drinkers who worshipped John Wilkes Booth. Now he offered the barn to Booth, and he groomed Booth's horses and kept them fed. Ned Spangler's biggest recompense came when Booth offered to buy him a drink at Taltavul's, next door to the theater.

As the men trotted out of town, the Washington City newspapers announced that President Lincoln would visit the Soldiers' Home, where Mr. and Mrs. Lincoln spent considerable time each summer, to witness a matinee of *Still Waters Run Deep*. At the Home, Dr. A. F. Sheldon, medical director, was busy supervising the dressing of the stage and the arrangement of chairs. He put on as many shows as he could, and was an early believer in recreation for ambulatory patients.

At 2 P.M. six men on horseback were waiting in a grove of trees a short distance beyond Seventh Street and Florida Avenue. The day was cold and gray and Booth explained, for the final time, the duties of each man. When the carriage was sighted coming around the bend, he and Surratt would ride out to meet it. They would assume a position ahead of the carriage and permit it to catch up with them.

The others were to wait a moment or two, then move out behind the carriage and not try to overtake it until they saw Surratt grab a bridle and pull it to a stop. The others were then to come up at once, and Paine would leap inside the carriage to subdue the President. Surratt and Booth would eliminate the coachman. Surratt would then don the driver's coat and silk hat and would drive the carriage. Booth and Paine would sit in back of the carriage with Lincoln. Arnold and O'Laughlin were to take care of mounted guards, if any. Atzerodt was to remain a few yards behind the scene, ready to respond to a call for help from any of the others.

The conspirators waited. There was no sound on the road. It was empty. The horses snorted and shook their heads. The men conversed in whispers. Carriage wheels were heard. Booth whispered, "How many?" but no one answered. He dashed out into the road to peer around the bend for an advance look, and came back shouting that this was the right one. In a moment, the others could see the shiny black vehicle coming around the sandy road.

The actor and John Surratt moved out. The others waited in concealment. Booth and Surratt rode ahead, then slowed, and, as the carriage came up, they parted so that they flanked the horses. Booth pulled rein and bent low to peer inside. There was one passenger, a smooth-faced man who looked startled. The actor motioned to Surratt to break away and the two rode back to the group. A violent argument broke out. Arnold and Surratt maintained that it was the President's carriage and the fact that the President wasn't in it proved that the government was aware of the plot and had sent the coach as a decoy. Federal cavalry would be along at any minute. Booth argued that this was merely the first coach to come along; that, if they had patience, Lincoln would be along in a few minutes. The men decided to wait.

In fifteen minutes, the conspirators broke up in fear and cursing. Lincoln had not shown up. Arnold and O'Laughlin swung away and, at a dead run, headed toward Baltimore. Atzerodt, with coattails flying, said that if anyone wanted him he would be at Port Tobacco and that he would stop at Surrattsville and explain everything to Davey. Surratt, angry, rode away alone. Booth and Paine, the one blind with rage, the other impassive, turned north toward Soldiers' Home.

The band would never again be at full strength. Arnold and O'Laughlin promised each other not to have anything further to do with plots. Surratt quit in disgust, because he had worked hard and earnestly for the Confederacy and he felt that this was an *opéra bouffe* plot.

Outside of Soldiers' Home, an actor named Edward Davenport was taking a breath of air when Booth rode up alone. The younger of the great Booths was wearing riding breeches, polished boots, and fawn-colored gauntlets.

"Good evening, Ned," said Booth. "Who is in the house today?"

"Hello, Wilkes," said Davenport. "Well, it is filled. Seward, Stanton, Chase—full up."

"Did the old man come?"

"The President?"

"Yes."

"No."

Booth turned away.

"What's the hurry?" said Davenport.

"I have a skittish horse."

On H Street, Mrs. Surratt was weeping. She sent Wiechman downstairs to early supper alone. The big man saw the tears and asked what the matter was. She shook her head. At the table, Wiechman told Dan, the Negro houseman, that Miss Mary was upstairs crying. Dan said that he knew that she was crying and he had asked her what was the matter and Miss Mary had told him she saw John go off on horseback with some other men and she did not like it.

At 6:30, Surratt came into the second-floor sitting room—the one immediately off the inverted V front steps—with a

Sharp pistol in his hand. Wiechman looked up from a newspaper and John waved the pistol angrily.

"My prospects are gone," he growled. "My hopes are blighted. I want something to do. Can you get me a clerkship?"

Before Wiechman could reply, Surratt was on his way upstairs to his room. A few minutes later, Paine came in with a pistol in his hand, saw Wiechman, and said nothing. He was breathing hard, like a man who had been running. Booth came in, talking loudly about his poor luck, and he paced the floor in agitation before he noticed Wiechman.

"I did not see you," he said, and went upstairs with the others. An hour later, Booth left the house, bound for New York, and Paine left for Baltimore.

On that day, President Lincoln had no appointment after 1 P.M. He planned to be at the matinee, and none of the patients at Soldiers' Home needed the relaxation more than he. He was ready to leave when Governor Oliver P. Morton, an imposing-looking man with arched brows and full black beard, walked in and said that he had just learned that one of his own regiments—the 140th Indiana—was coming down Pennsylvania Avenue and wanted to present a captured Rebel banner to him, and would Mr. Lincoln come along. The President reluctantly agreed.

So, at 2 P.M., while Booth and his band waited on Seventh Street road, the President was standing on the front steps of Booth's hotel, the National, telling an array of soldiers at parade rest that the war would soon be over and that they could then return to their families with the fervent thanks of the whole country.

On the morning of Saturday, March 25, two things of small moment happened. John Wilkes Booth returned to Washington and took a room at the National Hotel. John Surratt and a Mrs. Slater rented a buggy and drove off from the H Street boarding house. The widow Surratt said later that "John has gone to Richmond with Mrs. Slater to get a clerkship."

A notice appeared in the Washington *Star* on Monday, March 27, saying that the President and his family had reserved boxes at Ford's Theatre for the Wednesday night performance of the Italian opera *Ernani*. Booth saw the notice and wired O'Laughlin to come to Washington on Wednesday with or without Sam Arnold.

WE SELL THAT DAY SURE. DO NOT FAIL.

Neither showed up. Arnold had just learned that Wilkes Booth had stopped off in Baltimore last week, and had not even tried to find him. So Sam wrote a letter about it.

Hookstown, Balto. Co. March 27, 1865
Dear John:
     Was business so important that you could not remain in Balto. till I saw you? I came in as soon as I could, but found you had gone to W——n. I called also to see Mike, but learned from his mother he had gone out with you. . . . How inconsiderate you have been! When I left you, you stated that we would not meet in a month or so. Therefore, I made application for employment, an answer to which I shall receive during the week. I told my parents I had ceased with you. Can I, then, under existing circumstances, come as you request? You know full well that the G——t suspicions something is going on there; therefore, the undertaking is becoming more complicated. Why not, for the present, desist, for various reasons, which, if you look into, you can readily see, without my making any mention thereof. You, nor any one, can censure

me for my present course. You have been its cause, for how can I come after telling them I had left you? Suspicion rests upon me now from my whole family and even parties in the country. I will be compelled to leave home any how, and how soon I care not. None, no not one, were more in favor of the enterprise than myself, and to-day would be there, had you not done as you have—by this I mean, manner of proceeding. I am, as you well know, in need. I am, you may say, in rags, whereas to-day I ought to be well clothed. I do not feel right stalking about with[out] means, and more from appearances a beggar. I feel my dependence; but even all this would and was forgotten, for I was one with you. Time more propitious will arrive yet. Do not act rashly or in haste. I would prefer your first query, "Go and see how it will be taken at R———d" and ere long I shall be better prepared to again be with you. I dislike writing; would sooner verbally make known my views; yet your non-writing causes me thus to proceed.

Do not in anger peruse this. Weigh all I have said, and, as a rational man and a *friend,* you can not censure or upbraid my conduct. I sincerely trust this, nor aught else that shall or may occur, will ever be an obstacle to obliterate our former friendship and attachment. Write me to Balto., as I expect to be in about Wednesday or Thursday; or, if you can possibly come on, I will Tuesday meet you in Balto., at B———.

Ever I subscribe myself, Your friend,

Sam.

Booth wrote no reply. Arnold was out of the conspiracy, and he needed money. A second letter was delivered at the same time. This was from John Wilkes Booth's mother.

My dear Boy:

I have got yours. I was very glad to hear from you. I did part from you sadly, and still feel sad, very much so. June* has just left me. He staid as long as he could. Rose† has not returned yet. I am miserable enough. I have never doubted your love and devotion to me; in fact I always give you praise for being the fondest of all my boys, but since you leave me to grief I must doubt it. I am no Roman mother. I love my dear ones before country or anything else. Heaven guard you, is my constant prayer.

Your loving mother,

M. A. Booth.

The words "since you leave me to grief" cause one to ask "What grief?" Wilkes planned no known venture which could cause a mother to write about grief. The ensuing thoughts: "I am no Roman mother. I love my dear ones before country . . ." almost sound as though Mrs. Booth knew, or suspected, the plot. If she did, who in the family could have told her? Only Asia Booth Clarke, now pregnant in Philadelphia, could have known. It was with her that Wilkes left a letter to be made public in case of capture or death; the letter blamed Lincoln for all the woes of the South, and closed with "A Confederate doing duty on his own."

The end of March and the early days of April were dull for the conspirators. Lewis Paine—"The Reverend Mr. Wood"— was no longer welcome at the H Street boardinghouse and he checked into Herndon House at the corner of Ninth and F Streets. Mrs. Martha Murray, wife of the owner, gave him a big corner room on the third floor. She asked him if he would

---

* Elder son Junius.
† Daughter Rosalie.

also take his meals at Herndon House and he said yes. Dinner, she said, was served at 4 P.M. promptly.

The conspiracy was almost dead. Not quite, but close to it. Surratt had quit and had gone south, not for the clerkship he told his mother about, but to contact Mr. Judah Benjamin and to get a job sneaking Confederate dispatches to Canada. Sam Arnold was out of it too. He was getting a job in a grocery store outside Fortress Monroe, Virginia. Mike O'Laughlin was finished with plotting, although he was trying to achieve the delicate balance of not answering Booth's summons while keeping his friendship.

What was left after successive failures was Booth, Paine, Atzerodt and Herold. A brilliant actor, a stupid killer, a drunkard and a boy. That was the conspiracy, two weeks before the big day. Worse, Booth found himself out of funds. He took a train to New York, to borrow from his family and his friends. He saw Sam Chester and asked for the fifty he owed. He would not be asking it, Wilkes said, except that he was out of money and had to sell his horses. The ghost of the spurned plot was brought out, and Booth volunteered the information that, on inauguration day, he was as close to Lincoln "as I am to you." He could have shot the President, and now regretted that he had not.

These setbacks, poignant as they may have been to the conspirator, should have been as nothing compared to the one awful fact that there was no longer a Confederacy. It was, at this moment, in its death throes. In a trice, Mr. Booth had no cause, no country, no one to bring Lincoln to. In a day or two, the President would be in the place where Booth tried so desperately to bring him—Richmond.

Except that he would be there as a conqueror.

At 9 P.M. on Monday, April 3, John T. Holahan, tombstone carver and boarder at the Surratt House, was in bed. He wasn't close to sleep when he heard a soft rap on the bedroom door. He got up, pulled trousers over his nightshirt, went to the door and said: "Who is it?"

It was John Surratt. He had just come home on the Leonardtown stage from Richmond. He had left three days before, with an assignment which would take him to Montreal, Canada. He had been assured by Confederate authorities that, no matter what he heard in the North to the contrary, Richmond would not fall and the South would not surrender. He believed this.

"What is it?" said Mr. Holahan.

"John," said Surratt. "I would like to have some money."

Holahan opened the door and came outside.

"How much do you want?"

"Fifty dollars."

"You can have it. Wait here."

Holahan went inside and got the money. He counted it out. "Now, is that enough, John?"

Surratt studied the money in his hand and said: "I would like to have ten dollars more, making sixty in all."

Holahan went back into the room and got ten more. When he gave it to the landlady's son, Surratt said: "Take these," and gave Holahan two twenty-dollar gold pieces.

"I don't want them," Holahan said. "You can keep them. You are good enough to me for that amount of money."

"No," said John. "I want you to have them." Holahan took the forty dollars in gold for the sixty in paper.

That night, Surratt left for Canada. Train travel was slow and tedious; connections were poor. Whatever the specifics of his mission were has never been proved. It has been surmised that (1) his work had something to do with warning the Confederate group in Canada to flee (if so, this would be a sealed message and he would have no personal knowledge of it); (2) his mission had something to do with trying to free Confederate prisoners held in northern New York State.

At almost the same time that Surratt left Washington City, the clerk of the Aquidneck Hotel in New York City was presenting the register to an imposing young man and a pretty girl.

"J. W. Booth & Lady," the young man wrote. "Boston."

The couple was assigned to room number 3. Less than a month before, Booth had sat at dawn on a hotel bed with another girl, and had written on the back of an envelope:

Now in this hour that we part,
I will ask to be forgotten *never*
But, in thy pure and guileless heart
Consider me thy friend dear Eva.

Underneath, with whatever sad tenderness a dawn may be tinged, Eva had scrawled:

For of all sad words from tongue or pen—the saddest
are these—it might have been. March 5, 1865, in
John's room.

The morning of Tuesday, April 11, was rainy. The nation of towns and villages was still celebrating, still drinking and snapping its galluses over the surrender of General Robert E. Lee. On this night, the official celebration of the surrender would take place in Washington, but Mrs. Mary Surratt had no heart for it.

"Mr. Wiechman," she said, "won't you go around to the National Hotel and tell Mr. Booth that I sent you for his horse and buggy, and desire to know whether I can have it?"

At the National, Wiechman parroted the message and Booth shook his head.

"I have sold the horse and buggy," he said. Then he reached into his pocket. "But here are ten dollars. Go you and hire one."

Wiechman recalled that John Surratt had boasted that he owned the horses. "I thought they were John's horses," he said.

"No," said Booth. "They were my horses."

Wiechman accepted the money and walked in the rain up Seventh Street to Howard's Stable. He rented a high narrow buggy, with an oiled-cloth top, and a black horse. At 9:30 A.M. he chirruped to the horse and he and Mrs. Surratt started off for Surrattsville. It was a brisk drive—ten miles to Surrattsville, and ten home—and Mrs. Surratt always referred to it as "going down to the country."

This, to the widow, was just one more trip in an effort to reconcile her financial troubles. She was going to try to collect $479 and thirteen years of interest from a Mr. John Nothey. Long before, he had purchased seventy-five acres of farmland behind the tavern from her late husband. She had tried, by patience, by tact, by mail, to collect this money because one of her creditors was threatening suit. Now she was off to see Mr. Nothey and she hoped to come to some agreement with him before returning to Washington City tonight. In addition, she carried a small message from Booth to John Lloyd to have the "things" ready or the "guns" ready (the testimony varies) to be picked up.

It is this incidental message which makes Mrs. Surratt's trip so important because, three months from this day, she was going to be the first woman hanged in the United States and she would be hanged largely on the spirit and context of the message to Lloyd. The government would declare that this message proved that she was part of the John Wilkes Booth conspiracy—an active party to it. And she would go to her death denying that she was part of it; denying that anyone, including her son, told her about the conspiracy, and denying, in the presence of a priest, that she had been told to tell Lloyd anything about "guns." She was told solely to advise him to have the "things" ready. She was told, she said, that Lloyd would know what things.

The road began to dry on the Maryland shore. The skies were clearing and fresh as the buggy reached the rise of a hill. There was little conversation. The widow was angry with Mr. Nothey and she promised herself that she would squeeze him for every penny that was due.

At 11 A.M. the buggy was passing a crossroads called Uniontown when Mrs. Surratt noticed a rig passing in the other direction.

"Mr. Lloyd!" she yelled. "Oh, Mr. Lloyd!"

Wiechman pulled to a stop and, about a dozen yards behind them, John Lloyd pulled to the opposite side of the road. In his buggy sat Mrs. Emma Offutt, his sister-in-law, and her child. Lloyd got out of his buggy and walked across to Mrs. Surratt. The tavern tenant was sober and his hair was slicked. He smiled like a man who isn't sure that he won't regret it. He said he was going into town to make purchases and the widow said that she was going to the tavern to see Mr. Nothey. Lloyd said that he had seen Nothey around the tavern a few days ago, but wasn't sure that he was around today.

In a low voice (according to Lloyd) she told him to have the guns ready. "They will be needed soon," she said. Wiechman tried to listen, but said later that her voice was so low that he could not understand the words.

"I heard that the house is going to be searched," Lloyd said. "I do not like this, Mrs. Surratt." (This too is Lloyd's version of the conversation, and it was first quoted *after* his imprisonment.)

They chatted for a moment. "John has gone away," she said. "He will not be back for a while." Lloyd said that he had heard gossip that the government was about to arrest John for going to Richmond while it was still under siege.

The widow laughed. "Anyone in these days who can get to Richmond and back in six days must be smart indeed." The inference was that the government would have a hard time trying to prove that it could be done. She waved to the other buggy and shouted greetings to Mrs. Offutt, and resumed her journey.

They arrived at the tavern around noon and Mrs. Surratt became excited and shrill when she learned from the bartender that Mr. Nothey was not in the neighborhood.

"Please send someone," she said, "to fetch Mr. Nothey."

She and Wiechman drove to Bryantown for dinner. When they got back, Nothey was waiting in the front parlor. There was a private conference with the widow, and, late in the afternoon, Wiechman drove her home.

It seems impossible, almost a century later, to pin down the truth about this trip. Within a few days, most of these parties would be under arrest and those in charge of questioning them would be free with threats of death unless they "cooperated." Wiechman, a self-admitted coward, would strain himself to build up a case against the widow who befriended him; Lloyd, an alcoholic, would be told that he would hang with Mrs. Surratt unless his memory improved. It was his testimony which would send her to the gallows, although, within two years, he would recant and admit that he did not know whether she said "guns" or "things."

The next day, Wednesday, April 12, it was John Deery who noticed a change in Booth's attitude. Mr. Deery owned a saloon on E Street where it melts into Pennsylvania Avenue. It was directly over Grover's Theatre and it attracted the theater crowd plus devotees of cue and chalk because Mr. Deery was national billiards champion.

On this day, John Wilkes Booth stopped in before noon, asked for a bottle of brandy and water, and Deery remarked to himself that he did not remember the actor ever having done so much drinking as in the past few days. Also, Booth had never been so uncommunicative.

Deery polished glasses and tried to engage his old friend in conversation. Had Mr. Booth noticed, he asked, that the city council had been goaded into ordering a grand illumination of its own? All of the victory celebrations had been undertaken by the Federal Government and now the people had demanded that Mayor Wallach do something on a city level, and the council had decreed that tomorrow night—Thursday—would be their big night.

John Wilkes Booth looked up from the bar. He was dark and melancholy. Yes, he said, he had noticed. He wondered if he could have a little more water. Deery gave it to him. That closed the conversation.

Thursday the 13th was a picture-postcard day. The sun was yellow and billions of buds laced the trees in outrageous chartreuse. Tulip beds along the Mall began to display colored chalices. Forsythia showed graceful yellow everywhere and spun the buds on the breeze. The black earth cracked, and robins, fighting for space in an elm, cared not who lived in the White House. It was a clean good day, a day on which a warm breeze from the south warmed the cold stone of the Washington Monument and congealed the mud in the roads.

It was a nothing day to Booth until he learned that General and Mrs. Ulysses S. Grant were in Washington City. The news electrified him. The hero of heroes was in town and, Booth knew, the least that the Lincolns could do would be to invite him to stay at the White House and stage a state ball or a theater party for him. The heart of the arch-conspirator must have bounded with joy, because this news was custom-made for an assassin. Somehow, somewhere, there would have to be a public appearance of the President and the Man Who

Won the War. Booth asked nothing more than to learn when, and where.

Gossip was common that the Lincolns avoided state receptions because Congress had complained about White House expenses, and Mrs. Lincoln had huffily changed to the inexpensive theater party.

Booth reasoned that, if there was going to be a theater party, it would be held at Ford's or at Grover's. These drew most of the presidential patronage. The actor looked over the bills for the week, and counted on Grover's Theatre because they were opening with *Aladdin, or the Wonderful Lamp,* a new and dramatic vehicle, whereas Ford's had scheduled the old Laura Keene comedy, *Our American Cousin.* This play, he recollected, had not been well received even when it was new.

Booth walked up the Avenue to Grover's Theatre and, when he got inside, the theater was in cool darkness except the stage, where overhead lamps were lit. Onstage, Mr. C. Dwight Hess, manager, sat marking cue lines on a script with the prompter. In most circumstances, Booth had enormous respect for the theatrical proprieties and would not intrude on a script reading, but, on this occasion, he got up onstage and drew a chair.

The actor asked if Hess planned to join the city illumination tonight. The manager marked a place on the script with his finger, looked up, and said yes, to a degree, but that tomorrow night would be the big one as far as Grover's was concerned.

"Tomorrow?" said Booth.

"Yes. It is the fourth anniversary of the fall of Fort Sumter and tomorrow they're going to raise the flag over the fort again."

"Are you going to invite the President?" said Booth.

"Yes," said Hess. He shook his head. "That reminds me. I must send an invitation." (It was sent within the hour and was addressed to Mrs. Lincoln.)

Booth left and went upstairs to Deery's place. This time he was friendly and conversational. He said that he had seen Hess and that Grover's was going to stage a special celebration tomorrow in honor of Fort Sumter. Would Deery reserve the front right-hand box for Booth?

Deery chuckled. Why would one of the country's leading actors need a tavernkeeper to get box seats for him?

Because, said Booth, if I ask for it at the box office, Hess will feel impelled to extend the courtesy of the house, and I want to pay for this.

Oh, said Deery, in that case I can get them for you. He didn't blame Booth for not wanting to miss the show because, as he understood it, Hess planned to have a display of fireworks out front before curtain time, plus a Grand Oriental Spectacle, and a reading of Major French's new poem, "The Flag of Sumter." This, in addition to a performance of *Aladdin,* would make it a great evening.

The night before April 14 was cool and starry. John Wilkes Booth was on a rented horse, riding around town contacting Lewis Paine, David Herold and George Atzerodt. These were all that was left of the band. The carriage maker, weakest of the group, was at Pennsylvania House, a four-and-five-men-to-a-room hotel on C Street near Sixth. Wilkes ordered him to take a room at Kirkwood House, on the Avenue, and to spy on Vice President Andrew Johnson, who had a two-room suite in the first corridor behind the lobby. To each of his three men, Booth said that the time for action was at hand, and that this time there would be no failure because he planned to eliminate the President entirely.

Atzerodt was the only one who showed shock. Paine's reaction was casual. Herold was thrilled to be a part of such a shattering event.

By 8 P.M. the temporary gas jets in the windows of City Hall were blazing, and crowds were attracted to the big candlelit sign before the YMCA:

GOD, GRANT, OUR COUNTRY, PEACE

At midnight, Secretary of War Stanton was recopying his draft for peace and, a little more than a mile to the east, Booth sat in Room 228 at the National Hotel, also with pen in hand, and wrote a final note to his mother:

Dearest Mother—
I know you expect a letter from me and am sure you will hardly forgive me. But indeed I have had nothing to write about. Everything is dull, that is, has been until last night. Everything was bright and splendid. More so in my eyes if it had been a display in a nobler cause. But so goes the world. Might makes right. I only drop you these few lines to let you know that I am well and to say I have not heard from you. Excuse brevity; am in haste. Had one from Rose. With best love to you all.
I am your affectionate son, ever

John

Booth sealed it and prepared for the last good night's rest he would have.

The Morning Hours

9 a.m.

Mr. Lincoln folded the newspapers and put them to one side for further search, if time permitted. He signed two documents. Then he nodded to the soldier, now standing inside the double door, to admit the first visitor.

Watching him, on this final morning, a person with prescience and a sense of history would have recalled a lot of things that Lincoln had said which would make it look as though the President had known this day was coming.

"I do not consider that I have ever accomplished anything without God," he had said, "and if it is His will that I must die by the hand of an assassin, I must be resigned. I must do my duty as I see it, and leave the rest to God."

In an aside to Harriet Beecher Stowe, author of *Uncle Tom's Cabin,* he had said: "Whichever way the war ends, I have the impression that I shall not last long after it is over."

No one who heard him could doubt that he was philosophical about being killed, when he said: "If I am killed, I can die but once; but to live in constant dread of it is to die over and over again." To reinforce this point, one of the pigeonholes in his desk had a bulky envelope. It was labeled "Assassination" and it contained eighty threats on his life.

Nor did he worry about whether he was held in high esteem or low when he died. "I'll do the very best I know how," he had said, "the very best I can, and I mean to keep doing so until the end. If the end brings me out all right, what is said against me won't amount to anything. If the end brings me out wrong, ten angels swearing I was right will make no difference."

They had called him, among many other things, a nigger lover and, in a merciful fatherly way, he was. And he was the man to write: "As I would not be a *slave,* so I would not be a *master.*" But once, long ago, in the dancing heat of an Illinois summer, he had brought the same thought home to the heart: "When I see strong hands sowing, reaping and threshing wheat into bread, I cannot refrain from wishing and believing that those hands, some day, in God's good time, shall own the mouth they feed."

He had had his say on many subjects, and once, when a Christian minister had written that it was not right for the President of the United States to attend a theater when the nation was drenched in blood, Mr. Lincoln had written:

"Some think I do wrong to go to the opera and the theater, but it rests me. I love to be alone and yet to be with other people. I want to get this burden off; to change the current of my thoughts. A hearty laugh relieves me, and I seem better able after it to bear my cross."

The previous June, in Philadelphia, he had noted that many of the nation's newspapers were demanding Peace Now. And he had said, at a public gathering:

"War, at its best, is terrible, and this war of ours, in its magnitude and in its duration, is one of the most terrible. It has deranged business. . . . It has destroyed property, and ruined homes; it has produced a national debt and taxation unprecedented. . . . It has carried mourning to almost every home, until it can almost be said that the 'heavens are hung in black.' . . . We accepted this war for an object, a worthy object, and the war will end when the object is attained. Under God, I hope it never will end until that time. . . ."

The pale-eyed Speaker of the House, Schuyler Colfax, came in. He was a good-looking, brown-bearded man who was partial to long black coats and sleeves which exposed only his fingertips. He had the overly cordial manner and perpetual smile of the salesman. And what Mr. Colfax had to sell was always the same: himself.

They shook hands and Colfax sat. Lincoln liked him, not so much because no President can afford the enmity of the House Speaker, but rather because the President understood Mr. Colfax and appreciated him for what he was. Seven years before, Colfax had favored Douglas over Lincoln, but that had been forgotten. As a legislative leader, the man had worked fairly easily in harness with the executive and, even though the President knew that this man favored promotion over principles, and money over morals, he still looked upon him as a friendly rascal.

No notes were kept of this morning's discussion, but it is probable that they talked about Mr. Colfax's ambition to become a member of the Lincoln Cabinet. The President was not unstained in this situation. He had nourished the ambition in the Speaker's breast, and the conversation had reached the ways and means stage.

Everyone knew that Stanton had tried to quit as Secretary of War—a theatrical gesture, perhaps—and that Lincoln had thrown his arms around his favorite strong man and had begged him to stay on. But Stanton wanted to be appointed to the Supreme Court and, if the vacancy arose, there is no doubt that Lincoln would have presented his name to the Congress. This would leave the War Department open, and in a postwar period of demobilization and peace there would be no safer place to put Mr. Schuyler Colfax.

Another matter discussed was a growing congressional worry that Mr. Lincoln was about to undertake the reconstruction of the South without consulting the legislative branch of the government. Colfax tried to exact a presidential promise that postwar policy would not be laid down without calling a special session of Congress, but the best he could get from Mr. Lincoln was "I have no intention, at the moment," of calling a special session, but "if I change my mind, I will give the due sixty days notice."

As Speaker, Mr. Colfax was expected to protect the rights of Congress in this matter, but, as a Cabinet member persona-elect, he had to defend the position of the Chief Executive, who wanted to make a lenient peace with the South, with no outside help other than what he could expect from the Messrs. Seward, Welles and Stanton, and then only such help as he had to suffer.

This viewpoint was not a secret. Everybody knew that Mr. Lincoln wanted to go it alone and few, even among his friends, felt any sympathy for his viewpoint. Even those of his own party who agreed with him that a soft peace would be the most permanent peace felt obligated to speak up solemnly and admit that Congress was "entitled" to a voice in the matter.

Among the outspoken and bitter, Congressman George Julian said that Lincoln's views on reconstruction were "as distasteful as possible to radical Republicans." Wendell Phillips, still coining phrases, referred to Lincoln as a "first-rate second-rate man." Others, who had heard Lincoln promise the vote to the intelligent Negro and the Negro who had fought, murmured sadly: "Only those?"

When Colfax left, his step was springy and he beamed a greeting to those waiting outside the presidential office, so it could very well be that he felt he had lost the fight for Congress, but had a leg up on a Cabinet post for himself.

The next visitor was Congressman Cornelius Cole of California. What his business was is not known, but he remained overly long and the President wanted to close his morning appointment list before eleven because he had called a Cabinet meeting for that time. Two more men waited outside.

Upstairs, William H. Crook, the President's day guard, had relieved the night man in the corridor outside the Lincoln bedroom. He was a forthright, observant young man who had a honed sense of duty. On the 8 A.M.–4 P.M. shift, he was seldom more than a few paces away from the man whose life he guarded. Now, he took the night guard's hall chair and stuck it in a closet, and then turned off the gaslights in the hall.

He came downstairs, observing that the vultures had left, and checked the position of the military guards at the front gate, the front door of the Executive Mansion, and the door leading to the Executive Office.

The night guard, Alfonso Dunn, was standing on the front portico laughing at some of the victory celebrants who were reeling and roaring outside the White House fence. Crook joined him, half amused, and wrote later:

"Everybody is celebrating. The kind of celebration depends on the kind of person. It is merely a question of whether the intoxication is mental or physical. A stream of callers comes to congratulate the President, to tell how loyal they have been, and how they have always been sure that he would be victorious. The city is disorderly with men who are celebrating too hilariously." Later: "Those about the President lost somewhat of the feeling, usually present, that his life was not safe. It did not seem possible, now that the war was over . . . after President Lincoln had offered himself a target for Southern bullets in the streets of Richmond and had come out unscathed, there could be any danger."

John Wilkes Booth returned from the barbershop, smooth of skin and powdered. As he walked through the lobby of the National Hotel, the transients nudged each other and pointed him out. Sometimes people confused him with his brother Edwin, but this did not happen as often as it did in the early days. Wilkes was famous now; a star in his own right.

As he passed the desk, the clerk looked perfunctorily for the Booth mail. There was none. The actor's mail was usually delivered "c/o Ford's Opera House, 10th Street, Washington City, D.C." The Ford brothers could be trusted to hold all mail for him, no matter how long.

No one, not even the careful Mr. William H. Crook, would call Booth a suspicious person. Anyone who might have pointed an accusing finger at the noted actor on this particular morning would have been branded hysterical or insane. Everyone, it seemed, knew John Wilkes Booth and everyone, it seemed, admired him. He was the recipient of sunny smiles from strangers and, in his presence, demure ladies became bold. His name on the street billboards and fences was enough to guarantee a respectably full house, and Mr. Hess, the manager at Grover's Theatre, on E Street between Thirteenth and Fourteenth, would have given a great deal to get him away from the Fords.

Perhaps one person might have worried about Booth had he seen him this morning. His name was Oram, and years before he and Wilkes, as boys will, lay supinely in the grass at Bel Air and dreamed about the future.

"What I want," said Booth then, "is not to be so fine an actor as my father, but rather to be a name in history." He wanted, most of all, to be remembered. "I will make my name remembered by succeeding generations," he said, and the two had chewed long blades of seed grass and had watched the fat white dumplings of clouds float across a clearing between hummocks, and had spat green.

Another schoolmate joined them and he heard the story and he asked how it could be done. Wilkes had thought about the question and he had answered solemnly:

"I'll tell you what I mean. You have read about the Seven Wonders of the World? Well, we will take the Statue of Rhodes for an example. Suppose that statue was now standing and I should, by some means, overthrow it." The boys nodded. "My name would descend to posterity and never be forgotten, for it would be in all the histories of the times, and be read thousands of years after we are dead, and no matter how smart and good men we may be, we would never get our names in so many histories."

His friend Oram thought about it. Then he said: "Suppose the falling statue took you down with it? What good would all your glory do you?" And Wilkes propped himself on an elbow in the grass and smiled forgivingly. "I should die," he said, "with the satisfaction of knowing that I had done something never before accomplished by any other man, and something no other man would probably ever do."

There was no Oram around this morning. No one to feel other than envy at sight of Booth. As he walked back upstairs to his room, Michael O'Laughlin and his party of celebrants were coming down Pennsylvania Avenue. They had had breakfast, and some drinks, and hands and heads were steadier now. In fact, some of the gaiety of last night had been restored. The food at Welch's had been good, and the drinks, tasting stronger than yesterday, had recolored the city and made it a memorable sight.

There were four of them. Navy Ensign Henderson and Bernard J. Early were trying to agree on a mutual flat note with which to start a song. Mike O'Laughlin and Edward Murphy, leading the party, were silent. O'Laughlin was promising himself that he would not get drunk the second consecutive day, but he couldn't force himself to believe himself because he knew what a liar he was.

They were passing the National when Bernie Early said that, if nobody minded too much, he would pause and use the men's room. Nobody minded at all. Murphy, in fact, wondered if the hotel had a bar, and if not why not. They walked inside and O'Laughlin said that he had a friend in this hotel—John Wilkes Booth—and he would stop by and say hello to him in his room. Henderson and Murphy, while waiting, had *cartes de visite* made and, after forty-five minutes, sent two of them up by a hotel messenger to see if they could hurry Mike out of his friend's room. The cards came back with no answer, so the three went off to Lichau House, where men could sober up on pickled pigs' feet and big cold schooners of beer. Mike rejoined them there. He said that he had been trying to get some money that Booth owed him.

Over on the other side of town—the north side—Anna Ward had been to the post office early. She was young and unpretty and overly modest and nearsighted. She admired John Surratt very much. What hurt Anna's chances with John was that she already had the blessing of the boy's mother. This, as many girls have learned, can slow a romance to a walk. Still, John seemed to display fondness for her more when he was away from home than when he was in the boardinghouse. As today, for instance. She had two letters from him, both postmarked "Montreal, Canada East." Inside, in addition to the warm missives to her, were two enclosures for his mother.

Anna hurried as fast as she could to get to the boardinghouse on H Street to share her happiness with the widow Surratt. She had another letter for Mrs. Surratt postmarked Maryland, but Anna had no interest in it.

The two women, both nearsighted, read the letters aloud and hugged themselves over the news that John was well. He did not say what he was doing in Canada, but it was pretty well known that the Confederate States of America had shipped much money through Montreal to and from Europe, and it was tacitly understood that John Surratt had run the blockade for the Rebels between Richmond and Montreal, with documents concealed in a hollowed heel of a boot and between the floorboards of a rig. He had been stopped and searched several times by roving Union patrols in southern Maryland, but no one had ever found anything incriminating.

Now, the two women felt that the war was over, and that John would soon be home.

Mrs. Surratt read the Maryland letter by herself. It was from Charles Calvert, and he noted that he had tried, by every honorable means, to collect the money due him, and had failed. His lawyers had sought a judgment for the full amount in Maryland District Court. He would not have done this, Mr. Calvert said, except that he had met John Nothey, who had owed the Surratts $479 for thirteen years, and Mr. Nothey had said that he had tried to arrange a settlement of the debt, but that Mrs. Surratt did not seem interested in coming to terms. The little widow showed flashing indignation and said that, somehow, she would have to get to Surrattsville today. Mr. Wiechman usually drove her when John wasn't home, but the boarder was working and there was no man around to help. Still, she was determined to have a showdown with Mr. Nothey that day.

At the War Department office, immediately west of the White House, General Ulysses Simpson Grant was finishing his war's-end work. He stood, short and squat, tunic open, cigar in mouth, eyes squinting against the sight of the papers on the desk, and he approved the recommendations which would now go to Stanton.

The Secretary of War had been kind, in a curt, formal manner, to the hero of the Civil War. He had assigned a special office to him, saw to it that help was available, and had insisted that the general bring Mrs. Grant to the Stanton home. The secretary did not permit the general to forget that Stanton was boss but, other than that, he was somewhat pleased to find that the general was genuinely modest, did not want any honors, dreaded public appearances, asked for no voice in matters of policy, and wanted to visit his children in Burlington, New Jersey.

The secretary was surprised to find that, in Grant, there was nothing to guard against. He had asked the general, when he had come in from the front, to sketch a plan for cutting the size of the United States Army to something approaching a peacetime level, and to mark off which contracts for munitions and supplies could be canceled at once.

Grant had worked two days on the problem, often walking across the hall to the secretary's office for advice, going to the telegraph office to send dispatches, and digging into contract records. Now he had shown which divisions could be cut to what size (the Union Army would still have to patrol much of the South); and he had canceled contracts for items like shovels, ambulances, bayonets, cannonballs, food, uniforms, shoes. He found time to write two telegrams to General Meade, at Burke's Station, Virginia, granting permission to two Confederate messengers to return to Danville.

When the work was done, he went into Stanton's office and sat with the secretary. Mr. Stanton, fearful as always that President Lincoln would stumble, had sat up late last night working on his draft of proposals for peace in the South. He was still working on it. Grant told Stanton today, as he had on Thursday, of Mr. Lincoln's wish that the Grants accompany them to the theater tonight, but that neither one felt like attending. He did not tell something that Stanton may not have known at all: that Julia Dent Grant did not want to attend the theater tonight in the company of Mrs. Lincoln. In any case, the general said, he wanted to see his two children.

Stanton's silken whiskers hung over his papers and his head swung from side to side as he studied, first one draft of his peace proposals, then the other. He talked in short bursts and what he said was that he and Mrs. Stanton always turned down presidential invitations to the theater; they were, in fact, turning another one down this morning. His advice was not to feel bad; if the Grants did not go, the Lincolns would find someone else. The Cabinet members, he said, had all turned down such invitations and the Lincolns did not seem to be offended.

Grant felt better. He said that he was never much of a man for public appearances anyway, and he turned to his cipher operator, Samuel Beckwith, as though for confirmation. Mr. Stanton said that his advice to the general would be to wait until the Cabinet meeting at eleven, and then tell the President that he must decline. He added that Washington City was seething with intrigue, that, in some ways, it was as "Secesh" as Richmond, and that he had repeatedly asked Lincoln not to expose himself, either in theaters or at public gatherings.

The general and his cipher operator left the secretary's office and Grant sent a messenger to Mrs. Grant, at the Willard, to make arrangements to leave tonight for Burlington.

Lincoln did not know this.

10 a.m.

A convalescent sun came out, weak and pale, and young buds and rooftops glistened in the cool air. Ships which had been anchored downstream, loaded with prisoners, began to hoist anchor and sail and move upward against the southwesterly breeze.

In Charleston at this time of day, men stood proud and men stood sullen as they watched the Stars and Stripes lift jerkily up the halyard at Fort Sumter where, four years ago, it had come down. On the mainland, the Federal artillery belched fire and rattled the windows of the homes and the churches of Charleston.

An old alien authority had returned to South Carolina and the Union was going to have its celebration even if it had to do its own applauding. The slate blue smoke was lifting over the harbor when the Reverend Henry Ward Beecher spoke the speech that few Southerners believed—that the war was the fault of a small ruling class in the South, that the common people of the South would join their brethren of the North and rule the United States, that there was honor for all in this peace. Then he thanked God, who had sustained the life of the President, "under the unparalleled burdens and sufferings of four bloody years, and permitted him to behold this auspicious confirmation of that national unity for which he has waited with so much patience and fortitude. . . ."

In the White House, the man with so much patience saw a lawyer from Detroit, named William Alanson Howard. Then ex-Senator John P. Hale of New Hampshire, newly appointed to be Minister to Spain, sat down for a chat. He had served sixteen years in the United States Senate and he had permitted himself to become the biggest bone in a factional fight and had been defeated for reelection. Now he was grateful for an appointment to a $12,000-a-year job.

Nothing was said of his lame-duck status, and Lincoln was not the one to mention Hale's slashing attacks against the administration. The ex-Senator had confided to friends that, in one way, he was glad to leave the country because his daughter Bessie, he found, had succumbed to the blandishments of an actor named Wilkes Booth and he did not want an actor in the family; but that a sea voyage, and a long stay in Spain, would help the girl to forget the man.

Lincoln advised Hale to keep Assistant Secretary of State Frederick Seward informed of his preparations to leave, at least until the secretary was well enough to conduct the business of the department. The two men parted with a handshake.

The next visitor was shown into the office. For this man, the President arose from his big chair and came around to the other side of the desk for a warm handclasp. He was John A. J. Cresswell, the man who was credited with keeping Maryland from seceding from the Union. Lincoln, catching his mood from the broad beams of sunlight coming through the east windows, sat back in his chair and slapped both hands on the polished oak arms.

"Cresswell, old fellow," he said happily, "everything is bright this morning. The war is over. It has been a tough time, but we have lived it out"—then his voice dropped—"or some of us have."

In a moment, the notation of mass death was gone, and he said: "But it is over. We are going to have good times now, and a united country."

Cresswell agreed. The two men chatted about family welfare and the unique feeling of peace, and at last Mr. Cresswell got around to the favor he wanted to ask. It seems that a friend of his had gone south and, through some inadvertence, had enrolled in the Confederate Army. He had fallen into the hands of the Union Army and was now a prisoner. This friend had intended no harm, but there it was.

The President listened. Mr. Cresswell reached into his pocket and withdrew some papers. They were affidavits bearing on the friend's good character. "I know he acted like a fool," Mr. Cresswell said, "but he is my friend and a good fellow. Let him out, give him to me, and I will be responsible for him."

Lincoln's long fingers held a letter opener and he studied it as he turned the instrument in his hands. The ebullience of a few minutes ago was gone.

"Cresswell," he said, "you make me think of a lot of young folks who once started out Maying. To reach their destination, they had to cross a shallow stream, and did so by means of an old flat boat. When they came to return, they found to their dismay that the old scow had disappeared.

"They were in sore trouble and thought over all manner of devices for getting over the water, but without avail. After a time, one of the boys proposed that each fellow should pick up the girl he liked best and wade over with her. The masterly proposition was carried out, until all that were left upon the island was a little short chap and a great, long, gothic-built elderly lady. Now, Cresswell, you are trying to leave me in the same predicament. You fellows are all getting your own friends out of this scrape, and you will succeed in carrying off one after another until nobody but Jeff Davis and myself will be left on the island, and then I won't know what to do. How should I feel? How should I look lugging him over? I guess the way to avoid such an embarrassing situation is to let them all out at once."

Mr. Cresswell left, his mission unaccomplished. Afterward, other visitors came and left in rapid succession. Some sought pardons, one wanted a service discharge, others asked for approval to buy contraband in the South. The President wrote short notes directing department heads to take care of certain minor matters and, when two Southerners sent in word that they would like to have a pass to revisit Richmond, Mr. Lincoln sent a note out to them:

"No pass is necessary now to authorize anyone to go to and return from Petersburg and Richmond. People go and return as they did before the war."

In a momentary lull, he called for a messenger and asked him to run over to Tenth Street to Ford's Theatre and to tell the manager that the President would require the State Box for the evening performance and to explain that General Grant would be in the party. He probably forgot to mention Mrs. Grant because the newspaper accounts later in the day did not mention her.

Mr. Lincoln took care of this chore early for two reasons: one was that the impending Cabinet meeting might last a long time. The second was that he had a feeling that the last big Southern army, under General Joseph Johnston, would capitulate to General William T. Sherman this day, and he wanted very much to wait this news out either in his office or in the War Department telegraph office and this would leave him too busy to think of the theater.

Someone, perhaps Major Thomas T. Eckert, chief of the Telegraph Office, reminded Secretary Stanton that today was Good Friday. Mr. Stanton said that he was aware of the day. The subordinate said that many Christians, on this day, are in the habit of attending services or merely visiting churches. At once, Stanton wrote an order stating that all clerks under his jurisdiction, in whatever department, would be permitted to leave at once to attend services if they so desired. Copies of the order went out, and a carbon was signaled to all the forts and military installations in and around Washington. It would be left to the discretion of the commanding officer, in each case, to decide whether the individual could be spared from service for the day.

Across Long Bridge, two weary regiments trooped into Washington City and, without music or drums, marched down the Avenue. The men looked dusty. In ranks of four, they marched out of step, gawking at the city sights, the hotels, the restaurants, the taverns, the citizens standing on the north side of the Avenue, and especially at the patriotic women who waved handkerchiefs.

The boys were coming home from Virginia and the dirty blue columns would be marching day after day after day down the Avenue, until, in time, few pedestrians would pause to notice, and no handkerchiefs would flutter an engaging welcome. The men would still gawk, and spit tobacco juice, and make obscene remarks, but the only thing about them that would catch the warmth of light would be their bayonets.

The messenger reached Ford's Theatre at 10:30 A.M. and James R. Ford, business manager, was in the front office when he arrived. He heard the news, particularly about General Grant, with gratitude and enthusiasm. He sent the messenger back to the White House with word that the box was indeed available, that the Ford management was honored to have the President of the United States and his party as guests, and that suitable measures would be taken to entertain him.

Mr. James Ford couldn't stop smiling. Inside, a rehearsal was going on for *Our American Cousin,* and Ford had, until this moment, felt certain that Laura Keene and her company would be playing to an almost empty house on this final night of the engagement. Every theater manager knew that Easter week was the worst, for business, of the year. And the worst night of the worst week, by far, was Good Friday.

To make matters more discouraging for the Fords, their rival, Mr. Hess, had advertised that tonight Grover's Theatre would put on a monster victory celebration with lights and special songs. Ford had nothing to compete with it except an old comedy which, without Laura Keene's infusing fire, would have burned itself out long ago. Now suddenly, miraculously, the evening had been saved, and Ford realized that he must have Mrs. Lincoln to thank because, although the President had attended Ford's three times in the closing season, all three performances had been Shakespearean plays with Edwin Forrest. Ford could not imagine Mr. Lincoln *volunteering* to see *Our American Cousin*. It was not in character.

Still, a man has no right to question his own good fortune, and Mr. Ford bustled around the theater, passing the news to his younger brother and the actors on the stage. He stressed the fact that, while Lincoln was a fine attraction, General Grant would be the personage who would bring the crowd. Mr. Ford hurried out of the theater and over to the new Treasury building to get some bunting with which to decorate the State Box.

The exultation which James Ford felt is the more understandable because his older brother, Mr. John Ford, the owner of the theater was, at the time, in Richmond. For a week, James and young Harry Clay Ford were running the place. It was a coup to get the President of the United States on the poorest night of the year, but to get General Ulysses S. Grant was a major triumph.

John Ford, when he returned, would be pleased with his brothers. He was a veteran theater owner and manager, and had had houses in Baltimore and in Richmond. Once, by political accident, he had been Acting Mayor of Baltimore. He was known as a stubborn and independent businessman, but he was a family man too and, in spite of eleven children at home, he found time to worry about the safety of an aged uncle and a mother-in-law in Richmond. He had gone off to assure himself of their safety.

11 a.m.

The clocks were booming the hour when big Louis Wiechman, at work in the office of the Commissary General of Prisons, heard the news that Secretary Stanton would permit Christian churchgoers to attend services. At once, Wiechman asked for time off and, when it was granted, he asked if he must return to the office later. He was told that it would not be necessary to come back.

En route back to the Surratt boardinghouse, Louis Wiechman stopped in St. Matthew's Roman Catholic Church for a visit. This was at 15th and H, diagonally behind Seward's home, and a block away from the New York Avenue Presbyterian Church, where the Lincolns worshipped on occasion.

The President did not attend on this day. He sat at the small table between the big sunny windows chatting with the members of the Cabinet and greeting guests as they arrived for this important meeting. This one, he knew, would set the tone for the future of the South—and, by that token, the future of the North—and what was decided here today could hardly be thwarted by Congress before December, a cushion of eight months.

And so he was extra jovial in his greeting, standing to shake hands with a newcomer like General Grant, at which point the Cabinet members broke into applause. The two men sat near the window and chatted, while the other men broke into conversational groups. Frederick Seward came in, told the President that his father was improving slowly, then stepped away to permit Colonel Horace Porter to greet Mr. Lincoln.

The Secretary of the Navy, Mr. Welles, came in and sat in his Cabinet chair at the big table. He was puffing from the exertion of walking up the steps and down the long corridor. The Secretary of the Interior, John P. Usher, sat across from Welles and talked of spending a good part of the summer in Indiana. The newest member of the Cabinet, Hugh McCulloch, Secretary of the Treasury and the butt of Lincoln's gentle jokes about money, stood by himself at the start of his third Cabinet meeting, as though waiting to be asked to talk to someone. The curly-bearded snob from Ohio, Postmaster General William Dennison, did not ask him nor anyone else.

James Speed, the Attorney General, with a head like a ball anchor with moss dripping from the bottom, nodded to Mr. Lincoln and the President asked everyone to be seated at the Cabinet table. All hands were present except Secretary of State Seward, who was represented by his son; Vice President Andrew Johnson, who *might* have been invited but wasn't; and Secretary of War Stanton, who often arrived late.

The President sat almost sideways at the head of the table so that he could cross his legs, and opened the meeting by asking if there was any news of General Sherman. There was none, said General Grant softly, when he had left the War Department. No news from Sherman meant no news of surrender from General Joseph Johnston. The President said that he had a feeling that there would be news before the day was out.

He started to tell a story, in a soft, deprecating way, about a recurring dream he had had, and he had engaged the attention of his appointees when Mr. Stanton arrived, dropping his hat and coat in the anteroom, and apologized for being late. He said that he had hoped to bring great news of General Sherman, but he had none.

The story was not picked up again, and Lincoln channeled the talk toward reconstruction. Stanton, busy removing sheafs of papers from his portfolio, said that, unless there was disagreement, he would like to announce the cessation of the draft to the country. He felt that it would lift morale in all quarters, and help the country to realize that we all had to get back to work. Everyone was in agreement on this matter, and Stanton added that General Grant had spent a busy two days saving the country enormous sums of money by cutting the size of the army and by canceling army contracts.

The President said that it was pretty well agreed that the war was done, and that he needed suggestions regarding the best procedure to follow in the South. He was going to be saying those words, in different form, several times in the next three hours because he had no intention of adjourning this meeting for lunch or for any other reason until the skeleton outline of reconstruction had been erected.

Frederick Seward said that he had discussed this matter with his father before he left home this morning and, while it was extremely painful for his father to talk at all, the old gentleman had asked that the Cabinet consider ordering the Treasury Department to take charge of all Southern customs houses at once and to begin to collect revenues; he also felt that the War Department should garrison or destroy all Southern forts now; that the Navy Department, as a matter of precaution and a show of determination, should order armed vessels to drop anchor in all Southern ports and, at the same time, take possession of all navy yards, ships and ordnance; that the Department of the Interior should, without delay, send out Indian agents, pension agents, land agents and surveyors and set them to work reassessing Southern land; that the Post Office Department should reopen all post offices and reestablish all mail routes; that the Attorney General should busy himself with the appointing of proper judges and the reopening of the courts; in sum, that the Government of the United States should resume business in the South and that, at the same time, it should take care that constituted authority and private citizens were not molested or impeded in their tasks.

The President smiled. The man from the sickbed had said an enormous amount in one breath, and the dissection and digestion of each of these suggestions would require considerable agreement among the men around the table. Seward was aware of this, just as he was aware that the Secretary of War would be at this meeting with his own suggestions for picking up the broken states and reassembling them.

The meeting droned on. Mr. Lincoln listened, made suggestions, brought departments together in their thinking, and, when energies appeared to flag, he brought up another point in the Seward doctrine.

"We must reanimate the states," he said. He admitted to the gentlemen that he was relieved that the Congress was not in session and, by indirection, signified that he had no intention of calling it into special session.

Ideas caromed smartly around the green table, some clicking with all hands, some finding no proponents except the original sponsor. Communications would have to be re-established between the South and the North at once. Once the arteries were functioning, the next step would be to see that nourishment flowed through them and Mr. Stanton suggested that the Treasury Department be empowered to issue permits to all who wished to trade, and that he, at the War Department, order the Southern ports to receive all trading vessels.

Old Gideon Welles stroked his beard and thought that he saw Stanton reaching for postwar navy power.

"It would be better," he said, "if the President issued a proclamation stating the course to be pursued by each of the several departments."

This was an oblique warning to Stanton to leave naval matters to the Department of the Navy. Some of the men concentrated on wartime measures which could now be dropped. Stanton lifted his sheaf of papers and, raising his head slightly so that he could see through the lower part of his glasses, asked the Cabinet to listen to a plan which he had drafted "after a great deal of reflection" regarding reconstruction in the South.

He read the first part, pausing rarely to interpret, about the assertion of Federal authority in Virginia. Some of the members made penciled notes on pads set before them. When the reading of this section was finished, Mr. Welles reminded everyone present that the state of Virginia, which was used as a model in Mr. Stanton's document, already had a skeleton government and a governor.

The President, watching his flexing foot, said that the point was well taken. Stanton looked at the President to see whether the first section would be debated before he was allowed to proceed. Lincoln apparently was waiting to hear the rest of it. He had asked the Secretary of War to draw this thing up, and it was felt that the result would be a dilution of Lincoln's soft approach and Stanton's harsh one.

"I will see," said Stanton, "that each of you gets a copy of this."

The second section was then read and this part dealt with the reestablishment of the several state governments. Some of the members felt that there was more Stanton vengeance than Lincoln mercy in its proposals and Welles, as unofficial spokesman for the Stanton opposition, denounced it as "in conflict with the principles of self-government which I deem essential."

Mr. Stanton replied that he realized that the matter needed more work and more study, but that he had been asked to draw this up in time for the meeting and had done his best. Young Frederick Seward asked if it wouldn't be possible for all hands to be given a copy of the paper, so that it could be studied thoroughly and, at a subsequent Cabinet meeting, the President could have the best thinking of all members.

Mr. Lincoln nodded solemnly and said that he expected each man to "deliberate on this matter carefully."

Welles said that, as he saw the problem—taking Virginia as a hypothetical case—it was important, on the one hand, that the state government be representative of what the people of the Old Dominion wanted in a government, and yet equally impotant that the people be persuaded not to elect the very men who had instigated the rebellion in the first place.

Postmaster General Dennison said that Welles had stated the thing that he was worried about, and Mr. Stanton agreed. The sense of the members was that the best way to act would be to disenfranchise by name the leaders of the rebellion, political and military, and to leave the people of each state to choose from the men who were left.

Mr. Lincoln said little at this stage of the long meeting. He looked sadly and longingly at each man, and then looked back at his foot.

"We still acknowledge Pierpont as the legitimate governor of Virginia," said Mr. Welles.

"There will be little difficulty with Pierpont," said Dennison.

"None whatever," Mr. Stanton said.

Pierpont had assented to the breaking off of West Virginia as a separate state and the de facto recognition of Pierpont as governor of Virginia was involved.

Stanton blandly proposed that the states of North Carolina and Virginia be united under one state government, but the majority of the Cabinet members opposed the thought, claiming that the war had been fought to reunite the states, not to remake them.

Each messenger who tiptoed into the room was greeted with lighted expressions, but these dimmed quickly when it turned out that the news was not of General Johnston. There was no way that the men in that room could know that men in another room to the South were, at this moment, coming to a surrender decision. The messengers kept coming into Lincoln's big office, and each saw the same expectancy.

At one stage, in a lull, the President remembered his unfinished story about the recurring dream and he told the Cabinet that everything would turn out all right because of it.

"What kind of a dream was it?" said Mr. Welles.

"It relates to your element, the water," the President said. "I seemed to be in some indescribable vessel and I was moving with great rapidity toward an indefinite shore." Lincoln seemed to fear that the men might ridicule his dream, so he added: "I had this dream preceding Sumter, and Bull Run, Antietam, Gettysburg, Stone River, Vicksburg and Wilmington."

"Stone River was certainly no victory," said General Grant. "Nor can I think of any great results following it."

Lincoln agreed, but maintained that the dream usually presaged good news and great victories. "I had this strange dream again last night," he said, "and we shall, judging from the past, have great news very soon. I think it must be from Sherman. My thoughts are in that direction. . . ."

The talk turned to the leaders of the Confederacy and what to do with them. Mr. Stanton felt that, as Americans, they were traitors to their country and he could not see how anyone could put another interpretation on the matter.

Postmaster General Dennison, a roguish glint in his eyes, said: "I suppose, Mr. President, that you would not be sorry to have them escape out of the country?"

"Well," said Lincoln, leaning backward and trying to look serious, "I should not be sorry to have them out of the country, but I should be for following them up pretty close, to make sure of their going."

This led to an informal discussion of the many shades of opinion from the Senate and House on the subject of what to do with the leaders of the rebellion.

"I think it is providential," the President said, "that this great rebellion is crushed just as Congress has adjourned, and there are none of the disturbing elements of that body to hinder and embarrass us. If we are wise and discreet we shall reanimate the states and get their governments in successful operation, with order prevailing and the Union reestablished before Congress comes together in December. . . ."

That came from his head. This, from his heart:

"I hope that there will be no persecution, no bloody work after the war is over. No one need expect me to take any part in hanging or killing these men, even the worst of them. Frighten them out of the country, open the gates"—throwing up both arms and fluttering his fingers—"let down the bars, scare them off, enough lives have been sacrificed."

Wilkes Booth left his hotel. He stopped at the National Hotel desk and asked the clerk to have one of the boys post a letter for him. It was addressed to his mother, who was living with Edwin in New York. The clerk said that he would take care of it, and the young dandy walked out, turned up Pennsylvania Avenue, nodding pleasantly to the hero worshippers who nudged each other and gaped, and at Tenth he turned right and went up to Ford's Theatre. Here, he would pick up his mail. The letter he hoped to get would be from Sam Arnold, at Fort Monroe, agreeing to come to Washington at once.

The Ford Opera House (its formal name) was on the east side of Tenth Street between E and F. It was a three-story red brick building, now eighteen months old. It had five arched doorways, although only the one in the center led to the ticket office and the lobby. To the north of the theater—toward F Street—was Ferguson's Restaurant. To the south, the E Street side—was Taltavul's saloon. Up and down the block, on both sides, were narrow-front brick homes, each hardly more than twenty-five feet in width, standing shoulder to shoulder with the flanking houses. The street was unpaved and muddy, and a wooden ramp squatted in front of the center door of the theater so that ladies alighting from carriages would not get wet feet and soiled hems.

The actor stepped into the manager's office with the familiarity of an old friend. He picked up his mail, said hello to Harry Clay Ford. They were talking theatrical gossip when James J. Gifford, the stage carpenter, walked in and asked Mr. Ford what it was that he wanted. Ford said that President Lincoln and General Grant were coming to the theater tonight and to please have the stagehands remove the partition between Boxes 7 and 8. Booth masked his surprise. He was still smiling at the handwriting on the envelopes of his morning mail as he asked when the news was received.

A few minutes ago, Ford said. The President's messenger came in and said that, if the State Box was available, the President and General Grant and their ladies would use it tonight.

Had the President been invited?

No, as a matter of fact, he had not. In fact, nothing special except a new patriotic song would be on the bill tonight. After all, it was Good Friday and the last night of the Laura Keene company.

Mr. Booth nodded. He went out in front of the theater and sat down on the big gray granite step, his knees spread, his hands between them, reading his mail. At one point a passerby noticed him only because the actor was laughing heartily at something he was reading.

A few minutes later, Booth was no longer seen outside of Ford's Theatre. He was inside and upstairs.

The theater seated seventeen hundred persons and was gas lit. The patrons sat in straight-backed cane chairs which spread in curving rows and were split by two aisles, not counting the wall aisles. Tickets cost fifty cents and the management announced that there would be no extra charge for securing seats in advance.

Off the lobby was a double stairway which led up to the dress circle. This stairway was also used by patrons who had tickets for the upper boxes. The boxes were actually onstage; no part of them hung off into the orchestra section. There were two lowers on each side of the theater, and two uppers. These were designated by the management in the manner of theatrical directions, that is, from a position onstage looking toward the audience. Thus the State Box, or President's Box, was said to be at stage left, although, from the orchestra, it would be at the right-hand side.

This box was really two, numbered 7 and 8, both upper. These were separated by a partition of one-inch pine, held to the floor and ceiling by little L's of angle iron. When the President of the United States was going to use the box, the partition was always removed and the numbers 7 and 8 became one. As a matter of respect for the President, no other boxes in the house were sold on those occasions.

The doors to Boxes 7 and 8 were of three-eighth-inch pine, milled so that each had a large upper panel and a small lower one. Each had its own lock and its own key, although, when the Lincolns were present, only the door to the rear box, number 7, was used. Outside the boxes was a small unlighted hallway. Behind it was a small white door which faced the patrons in the dress circle. Outside this door, the President's guard sat on a chair, with his back toward the door, and facing the side aisle of the dress circle. Anyone approaching Boxes 7 and 8 would have to come down this aisle.

The stage itself was neither deep nor imposing and, in a scene of shallow proportions, a patron sitting forward in Box 8 often looked down toward the back of the head of the actor.

The distance between the ledge of the upper boxes and the floor of the stage was eleven feet.

Under the stage was a passageway in the form of a T. This was necessary to connect both sides of the stage and to leave a subterranean passage for the musicians who, although they had no orchestra pit, played their music in the space between the footlights and the front row of the orchestra. Also under the stage was the main gas valve, for dimming all the lights of the theater when the curtain was about to rise, and to brighten them between acts and at the end of the play. Gas jets in light-colored bowls were bracketed around the walls of the theater, and over the stage.

At stage right (left from the orchestra) was a small enclosure called the Green Room. Actors who were imminently due onstage came downstairs from their dressing rooms and sat here before cue time. To the rear of the Green Room was a door, leading to the back alley. On the far side of the alley were the backyards of houses which faced Ninth Street, and also up toward F Street a few Negro shanties with earthen floors and heavy curtains instead of front doors. At night, these were candle lit.

Booth walked around the back of the dress circle, toward the right, and down the steps of the side aisle to the little white door leading to the State Box corridor. He went inside and turned left through the door of Box 7—the rear box—and sat watching the rehearsal onstage.

He was acquainted with almost every line of *Our American Cousin*. In the box, with no gas lights on, he was cloaked in daytime gloom and he sat watching, thinking—who knows? One thing seems certain. He timed his plans for tonight now. He looked from the ledge of the box to the stage, and he knew that he had made bigger leaps in *Macbeth*. He could not plan to run back through the dress circle because, the moment the act was accomplished, it could be expected that the people in the theater would be in bedlam. Besides, he would have to stab the guard outside the little white door in case of challenge. It was better to stick to the original idea, to jump to the stage, run across toward the Green Room, and out the back door. If he had a horse there, waiting, escape should be fairly easy. That too, had been planned for a long time.

For a while, he sat watching the actors run through their lines lightly and with little feeling. They were in the second scene of the third act, the part where the English mother, Mrs. Mountchessington, in trying to marry her daughter off to Asa Trenchard, the rich American, first learns that he is not rich.

"I am aware, Mr. Trenchard," she says, outraged, "that you are not used to the manners of good society, and that alone will excuse the impertinence of which you have been guilty." She flounces offstage, leaving Asa Trenchard (Mr. Harry Hawk) alone on the boards. He watches her go, then grunts. "Don't know the manners of good society, eh?" he says. "Well, I guess I know enough to turn you inside out, old gal—you sockdologizing old man trap!"

This bit always drew one of the biggest laughs of the play. Booth watched and did not laugh. He was fascinated. At this point, the stage would not be cluttered with actors. Harry Hawk would be alone. Mrs. Muzzy, who played the haughty English mother, would be on her way to her dressing room. Laura Keene would be in the wings, waiting for a cue. Mr. Gifford and his two stagehands would be backstage storing the flats from the previous scene.

Harry Hawk was still talking: "Well now, when I think of what I've thrown away in hard cash . . ."

John Wilkes Booth was not listening. He was thinking. And what he was thinking of—if one can hazard a guess—was that if the curtain rose at 8 P.M. (and it usually did) then this particular scene should be on in about two hours or a little bit more; 10:15 perhaps.

The actor had seen enough of the rehearsal. He looked at the partition between the boxes, as yet unremoved, and he walked out into the little corridor, examining the doors, and left the theater.

There was a lot to be done, and precious little time in which to do it. Tonight, he would pull down the Colossus of Rhodes.

The Afternoon Hours

12 noon

At noon Washington City was quiet. The sun was obscured and the view was heavy with haze. Many of the government employees had taken advantage of Stanton's order, and similar orders in other departments, and had gone home. No church bells sounded. Few pedestrians were on the streets. It was twelve o'clock on Good Friday and this was the hour that Christ had been nailed to the cross.

It was an unnatural quiet, an uneasy quiet. The men at the long produce market on the south side of the Avenue worked at empty stalls, gutting shad, shucking oysters, butchers turned slabs of beef over in brine barrels, and all of them looked up and down the street and wondered what had become of the people.

The people were in church, or at home. They knelt, or they dozed. Even the bars were held erect by the very few and the very strong. Some honeymooners stopped at Gardner's studio to pose—he sitting, she standing—for a lifetime memento. At the foot of Fourteenth Street, the daily thunder of army wagons could be heard on the loose planking of Long Bridge, coming home from war.

James R. Ford, in a buggy, was returning from the Treasury Department, laden with flags for decorating the President's Box. He was walking his horse along E Street, and was turning off onto Tenth when he saw Booth. Ford pulled the horse up, and they chatted. It was a dull day, Ford admitted, but, with Grant and Lincoln present, the house was sure to be a sellout. Booth asked him if he had got all the flags he wanted and Ford said no, that he had asked Jones for a thirty-six-foot American flag that the Treasury used on special occasions, but that Captain Jones had told him that the flag had been on loan for the illumination and wasn't back. Ford wanted it to drape down the upper floors of the front of the theater.

The two men parted, Booth saying that he would try to attend if he could, but not promising. James got back to Ford's and, after getting his bunting indoors, asked an actor to write a special notice for the Washington *Star* and the *National Republican* announcing the presence of Grant and Lincoln as honored guests tonight. The actor said that it would have to wait; he was busy writing the regular advertisements.

James wrote the announcements himself, and then he worried about the propriety of it. Until it was on paper, it had seemed all right to capitalize on the presence of two great men. Now, as he read it back, it seemed cheap. He called young Harry and asked for an opinion. They read it aloud together and it seemed all right. They agreed that such an announcement could harm no one, and, at the same time, it was bound to draw the patronage of transients.

A colored boy delivered both by hand.

Wilkes Booth walked up to G Street, and across G to Seventh, and stepped into Howard's Stable. The stableman knew him, and Booth asked that his big one-eyed roan be delivered to the little stable behind Ford's Theatre and tethered there. He paid the feed bill for the horse, and left. Then he took the long walk down across the Mall to Pumphrey's Stable, and asked for a sorrel which he had been renting for six weeks past. The stableman said that the sorrel was out and Mr. Booth could not have him. Instead, he said, he had a fine, fast roan mare for hire, a little nervous perhaps, but a good fast riding horse.

The stableman brought her out and turned her around inside the door. Booth studied the animal. She was young, about fourteen hands high, and she had croup chafes on her quarters. Her mane and tail were black. She had one white sock and a star on her forehead. Booth liked her skittish bearing.

"Have her saddled at four o'clock," he said. "I'll be back."

George Atzerodt, new resident at Kirkwood House, Twelfth Street at the Avenue, was in and out of the hotel in the manner of a baggy-pants comic who knows that if he does something ridiculous three times it will induce laughter. On the hotel register, he had scrawled his right name: "G. A. Atzerodt" and, although he had not been in his room on the second floor since early morning, David Herold had been in it to leave clothes and weapons.

Atzerodt spent most of his time drinking at the bar and trying to be disarming. He asked so many questions of the bartender and the few customers that he excited suspicion. Where exactly, he wanted to know, is the Vice President's room? Does he have a guard? Could any citizen knock on the door and have a chat with him? A man in his position doesn't carry firearms, does he? How about the nigger who stands behind him when he eats—where does he go when Johnson goes back to his quarters? Does the Vice President stay home at night or does he go out? Would you say that he was a brave man or a coward? Ever see any soldiers around him?

The men at the bar studied Atzerodt, and the stupid face squeezed into a smile, and they relaxed. The glances around the bar were a shrug—this man was a drunken outlander.

Booth went back to the National Hotel to dress. He wanted calf boots and new spurs and a black suit with tight trousers, for riding. He wanted a good black hat too, and his wallet with the pictures of his girls, and the diary which had been bare of words. He wanted a small pocket compass, his gold timepiece, a small brass derringer, a gimlet and a long sheathed knife which could be stuck inside the trousers along the left side.

There would be no more failures; he knew that. Lincoln would die tonight, or Booth would. Or perhaps both. The last possibility did not frighten him. The actor was aware that the chances were that he would not get away, rather than that he would. Pulling the huge statue down with him was the important thing. The only thing.

If he worried, it was over his fellow conspirators. No one knew better than he that these men—his band of irregulars— were ciphers; nothings; buffoon assassins. A sniveling alcoholic, a giggling boy, and a brainless automaton. Atzerodt would be assigned to kill Vice President Johnson and Wilkes would be surprised if George approached the man at all. However, Johnson was the least important of the men he wanted to kill—a white trash tailor—and that's why Atzerodt was assigned to him.[*] Seward was more important, and Paine would really kill the man if . . . if . . . if Paine could find him.

In the past several weeks, Booth had tried to make Washington City comprehensible to Lewis Paine, but the ex-soldier became more confused. He could not understand uptown from downtown, north from south, or right from left. Like a hound dog, he would have to be taken to the bush the bird nested in, and pointed. Davey Herold would do that. So, in substance, he had to use two men to get one sick man killed, and a third man to get nothing done. Only he, John Wilkes Booth, had the courage, the intelligence, and the patriotism to walk in "among a thousand of his friends" and slay the Despot Himself.

The plan called for meeting afterward on the road to Surrattsville. And this too, the actor must have known, was a dream. Each of them would bungle and fail, and each would

---

[*] Whatever his other faults may have seemed, Johnson was an exceedingly tough customer, a man of very great physical courage, who had been threatened with assassination by experts all through the war. Actually murdering Johnson was about as tough an assignment as a conspirator could have been given—and of all people in Washington, Atzerodt got the job!

bring sudden death on himself. In cool assessment, Booth was pretty sure that he would not be on the road to Surrattsville tonight. Getting into the State Box was a formidable problem; getting out of the theater and away was going to be almost impossible.

For a while, he toyed with the idea of taking Paine with him to Ford's Theatre. Paine could be assigned to kill General Grant while Booth was dispatching President Lincoln. But the actor dropped the plan quickly. It would be infinitely more difficult to get two persons into that box tonight than one; and it would be more than twice as hard to get two people out of that theater, than one. Besides, if Paine assassinated Grant, and Booth failed to kill Lincoln, the actor would be a fool in history. His theatrical sense warned him not to share billing with anyone. He would do it himself—Lincoln with the gun; Grant with the knife.

Mrs. Surratt and Miss Honora Fitzpatrick left the boardinghouse and walked over to St. Patrick's Church, on Tenth Street between E and F, to pray during a part of the Three Hours Agony. The church was dim and cool. The Crucifix was covered with purple cloth; so were the statues of the Blessed Mother and Saint Joseph. And the Stations of the Cross which lined the walls of the church. Communicants knelt in pews and their lips moved in sibilant whispers. Their eyes blinked toward the now empty repository of the Holy Eucharist.

Others, seeking salvation, knelt in the Baptist Church, in New York Avenue Presbyterian Church, in the Methodist churches, the Episcopal. In Georgia, by the clock, He was dying; in Rome He was already dead; in California He was not yet on the cross.

1 p.m.

Young Mr. Harry Ford stopped next door at Ferguson's Restaurant for lunch. He saw big, ham-handed James P. Ferguson behind the cigar counter and he said: "Your favorite, General Grant, is going to be in the theater tonight. If you want to see him, you had better go get a seat."

Ferguson thanked him, asked a question or two, took off his white apron and ran next door to get a seat. In fact, he wanted two. There was a little girl who lived next door to him, and she showered a shy adoration on big Jim, and now, with her mother's permission, Ferguson would take her with him.

This man was sensitive to history and to historical personages. He had seen Mr. Lincoln many times, but he would still run out of his restaurant to watch him pass by in a carriage. However, he had never seen the Little Giant and tonight would be his opportunity. Mr. Maddox was in the box office and he tried to sell Ferguson two good seats downstairs, but the restaurateur told him that he wasn't going to see *Our American Cousin*; in fact, he wouldn't care if he never saw it. What he wanted to know was, will the President use the usual boxes—7 and 8? Maddox said he would. Then, said Ferguson, I want two front-row seats on the left-hand side of the dress circle, because that's the only place in the house with a view into the Presidential Box.

He got them.

In President Lincoln's office, the Cabinet meeting continued. It had passed its second hour, and the President was pleased to note that, except for minor differences of opinion his Cabinet seemed to be agreed that, if the South were helped to get on its economic feet, the effect would be to enhance the welfare of the North. No one, including Lincoln, desired to spoon-feed the South and, by the same token, no one wanted to heap additional punishment on the defeated states. Stanton was for a sterner peace than the President, but the difference between the men was not beyond bridging.

In and out of Congress and the newspapers there were varying shades of public opinion about this matter, ranging all the way from those who desired to re-embrace the South and start anew, to the bitterness of Senator Ben Wade, who hoped that the Negroes of the South would be goaded to insurrection, feeling that, "if they could contrive to slay one half of their oppressors, the other half would hold them in the highest respect and no doubt treat them with justice."

Horace Greeley, an editorial flirt, had been Lincoln's friend and was now his enemy. A year ago, he had parted politically from Lincoln when, in the New York *Tribune,* he had begged for peace at almost any price. Now he opposed Lincoln politically and personally.

After lunch, in New York, he went to the office of his managing editor, Sidney Howard Gray, and handed to him a sheaf of papers written in longhand. It was an editorial for tomorrow's paper and would be off the composing room floor at 2 A.M. Gray, accustomed to Greeley's attacks on the President, read it after the boss left and found it to be so "brutal, bitter, sarcastic and personal" that, though he had it set in type, he hid the galley.

The President was aware, on this day and at this meeting, that, in America, he was now a minority political leader. The entire South, temporarily disenfranchised, opposed him. The Democratic party of the North opposed him. The radicals in his own Republican party opposed him. Most of the influential newspapers opposed him. Even the mild Senator Morrill of Maine found it "truly most difficult to speak of the elements of Lincoln's character without offending public sense." He was scorned, maligned, spat upon as a person lacking decision, character, intelligence and honor. He was an ape, a buffoon, a rascal of dirty mind and dirty jokes. He was held in low esteem by politicians and molders of public opinion. The only persons who loved him were the people, and they would not fully realize it until tomorrow.

To all of which Mr. Lincoln said: "As a general rule, I abstain from reading the reports of attacks upon myself, wishing not to be provoked by that to which I cannot properly offer an answer. . . ."

That is why, at this meeting, he was determined to block out the form of the peace no matter how long he had to hold the Cabinet together. He did not expect it to be formalized by signed documents; he wanted basic agreements. If that could be achieved, and the machinery to implement it were set in motion, then he would consider that what he had started out to achieve in 1861 had been, in the main, realized.

He noted with satisfaction that when Secretary of the Navy Welles offered his views about the course to be followed in Virginia, Secretaries Stanton and Dennison agreed, and the others said nothing in opposition. Thus, even among men who distrusted each other, Lincoln had harmony on this one day.

Frederick Seward saw "visible relief and content" on the face of the President and said that, in the regular order of business, none could refrain from chatting about "the great news" of the war's sudden end. Like boys who had recovered from an interminable illness, they tried to stick to their schoolwork, but they could not refrain from looking out the window at the sunshine and the lush grass and the dizzying atmosphere of feeling good.

The President asked General Grant to tell the Cabinet the details of General Lee's surrender, and Grant did, detracting from his own role and saying nothing to lessen the figure of Lee. When he had finished, Mr. Lincoln spoke.

"What terms did you make for the common soldiers?"

Grant fingered his beard and said: "I told them to go back to their homes and families, and they would not be molested, if they did nothing more."

In front of Kirkwood House, Vice President Johnson, hands thrust in pockets, scowled at the unnatural quiet of the city and walked back into the hotel. He went to his rooms, closed the door, and sat reading.

A short time later, George Atzerodt came in with a bundle and walked up the curving staircase to his room. If he made any noise in his alcoholic anxiety, it is not recorded that the Vice President, in a room almost directly below, made any complaint. He deposited huge pistols under the pillow on his bed and a knife under the sheet. He gave passing glances at the coat and materials left by Herold, and then he went back downstairs, where he asked the room clerk to point out the Vice President's room, and asked where Johnson was right now.

The room was pointed out to him, and he was told that the Vice President had just come in. George Atzerodt's reaction was to straighten up in shocked surprise, and to step into the bar. He was there only a short time when he paid for his drinks and walked out.

A few minutes later, the Vice President, having scanned a newspaper, glanced at his timepiece and got up and put his coat on and left. He had an appointment, an after-lunch appointment, with the President. Neither of them had been sure how long the Cabinet meeting would take, so Lincoln had suggested that Johnson "drop over" early in the afternoon.

Johnson walked up Pennsylvania Avenue, across Fifteenth at the Treasury Building, and around the corner to the White House. At the gate, two soldier guards recognized him and snapped to attention. He nodded without smiling and walked on in. At the front door, he was met by a colored doorman, a soldier with carbine and bayonet, and the President's personal guard, Crook.

They held the door for him and he walked inside, down the corridor and up the staircase until he reached the big doors of the President's office. The soldier at that door said that he was sorry, but that the Cabinet meeting was still going on, and the President had not had his lunch. Mr. Johnson said that he would stroll around until the President was ready.

2 p.m.

The early editions of the afternoon newspapers were being hawked on the streets and, on page 1 of the *National Intelligencer*, the top half of column four was taken up with an advertisement from Grover's Theatre, which announced

THE

THE

THE

THE

GORGEOUS PLAY

GORGEOUS PLAY

GORGEOUS PLAY

GORGEOUS PLAY

OF

OF

OF

OF

ALADDIN

ALADDIN

ALADDIN

ALADDIN

OR

OR

OR

OR

THE WONDERFUL LAMP

THE WONDERFUL LAMP

THE WONDERFUL LAMP
THE WONDERFUL LAMP

Underneath this, in a small advertisement, Ford's Theatre announced the "Benefit and last appearance of Miss Laura Keene in her celebrated comedy of *Our American Cousin*." The Washington *Evening Star* was on the street with three small announcements, spread through the newspaper, which announced the presence of General Grant and Mr. and Mrs. Lincoln at Ford's Theatre this evening.

Under "City Items," the announcement said:

Ford's Theatre—"Honor To Our Soldiers." A new and patriotic song and chorus has been written by Mr. H. B. Phillips, and will be sung this evening by the Entire Company to do honor to Lieutenant General Grant and President Lincoln and Lady, who visit the Theatre in compliment to Miss Laura Keene, whose benefit and last appearance is announced in the bills of the day. The music of the above song is composed by Prof. W. Withers Jr.

A small advertisement in the *Star* pointed up the fact that the name of Mrs. Grant was not omitted by accident or by typographical error:

Lieut. Gen'l Grant, President and Mrs. Lincoln have secured the State Box at Ford's Theatre tonight, to witness Laura Keene's *American Cousin*.

Two hundred and fifty miles to the north, the *Whig Press* of Middletown, New York, published the news that President Lincoln had been assassinated. No one found out the sources of this "news," or why the editor published it without checking by telegraph with General John A. Dix, commandant in New York, who would be among the first to know any such momentous news, or with the President's secretary, John G. Nicolay, at the White House.

In this time of poor communications, it is puzzling that, at the same hour, the people of Manchester, New Hampshire, were buzzing with rumors that President Lincoln had been shot. In far-off St. Joseph, Minnesota, forty miles from the nearest telegraph office, men met other men on the street and asked, in horrified tones, if they had heard the news that President Lincoln had been killed by an assassin.

The Cabinet meeting was drawing to a close. The interested parties were in basic agreement and the President said that he guessed they would all have to wait until later to hear from General Sherman. He got to his feet, and pulled his heavy gold watch from his vest. The other men arose and the meeting was adjourned.

The general, with Colonel Porter a pace behind him, walked over to the President and thanked him for the opportunity of attending a Cabinet session. Lincoln said it was a particularly amicable one and he was glad the general could be present. Grant then brought up the matter of the theater party. He fidgeted and seemed embarrassed and he said that Mrs. Grant would be sorely disappointed if their visit to the children was delayed any further. In fact, Mrs. Grant planned to take the evening train to Philadelphia.

Lincoln, looking down on his hero, joshed the commanding general and said that there would be plenty of time to see the children, and reminded him that the people would get great pleasure out of seeing at firsthand the man who had won the war. Grant subsided. He had not said "I will not go." Nor had he followed Mr. Stanton's suggestion and said: "You should not go. Your life may be in danger."

The groups were still chatting in the room when Colonel Porter brought a note to Grant. It was from Mrs. Grant and it said that she hoped that the general would not delay their departure on the six o'clock train. This was exactly what Grant needed to face up to the President. He showed the note, and said that he *must decide* not to remain in Washington. The President understood that this was a final decision.

Contrary to report Stanton did not participate in this conversation and there is no record that he stood beside his general and said: "Neither of you should go to the theater tonight." As the meeting broke up, the President shook hands with Secretary of the Treasury Hugh McCulloch and said: "We must look to you, Mr. Secretary, for the money to pay off the soldiers."

"You should look to the people, Mr. President," said McCulloch.

The President smiled. "They have not failed us thus far," he said, "and I do not think they will now."

Frederick Seward bowed gravely and made his farewell. He reminded the President that a new British Minister, Sir Frederick Bruce, was in Washington awaiting his presentation to the President.

"Would any time tomorrow be convenient?" Seward said.

The President clasped his hands behind his back and studied the ceiling.

"Tomorrow at two o'clock?" he said.

"Yes," said Seward. "In the Blue Room?"

"Yes," Lincoln said. "The Blue Room."

He had a small boy's scorn of the spurious expressions of esteem engendered in a formal meeting between ambassador and chief of state and he particularly winced at the florid speeches which the State Department made him read to newly arrived ambassadors.

"Don't forget to send up the speech beforehand," he said. "I like to look it over."

As the men left the office, Grant asked Stanton if he might stop by at the office later and see him.

"I'll be there until evening," Stanton said.

Lincoln left the office at about 2:20 for lunch with Mrs. Lincoln. He did not see Vice President Johnson and, as the coquettish weather had become sunny and warm again, it may be that Johnson was walking around the grounds. There is no record of what conversation transpired at the lunch but it is almost certain that he told Mrs. Lincoln what he would say to others later in the day: Grant is not coming. I do not feel like going to the theater tonight and, if it were not for disappointing so many people, I wouldn't.

On H Street, Wiechman was writing a letter in his room. There was a knock. It was Mrs. Surratt. She said that she was sorry to intrude on his time, but that she had just received a letter from Mr. Charles Calvert and that Mr. Calvert said that she would have to go to the country again if she expected to collect her $479 from Mr. Nothey.

Wiechman folded his letter and said that he would drive her. She gave him ten dollars and asked him to get a horse and buggy. Just as Louis Wiechman was leaving the house (this was 2:30 P.M.), John Wilkes Booth walked in. The inquisitive boarder was burning to remain and listen, but he had no excuse to do so, and, as he turned to close the front door behind him, he looked around and saw Booth with his back to the marble fireplace, one arm flung across it. Mrs. Surratt stood facing him.

The big fellow hurried around the corner to Howard's Stable and got a horse and buggy as quickly as possible and trotted the horse back to the boardinghouse. He hitched the horse and started up the inverted V steps, just as Mrs. Surratt came out of the upstairs sitting room and said that, now that Booth had left, she was ready to go. He turned back to the buggy and she said: "Wait, Mr. Wiechman. I must get those things of Booth's."

"Those things of Booth's" turned out to be a small package, about the size of a saucer, covered with brown paper and tied two ways with string. She got into the carriage and put the package on the floor between her feet.

"I must be careful," she said, as Louis swung the horse out on H Street and started toward the Navy Yard Bridge. "It's brittle. Glass. I do not want it to get wet."

The government clerk waited to hear more, but the widow had nothing further to say. She talked about the uncertainty of the weather and of how a body hardly knew what to wear these days, with rain and chill and then a hot sun and no breeze, but that the nice time of year would be along in about a week and, down in the country, they had been plowing and planting for two weeks and they were lucky that there hadn't been a frost because, when Mr. Surratt was alive, she had learned what one frost can do, if it is a proper freezing frost.

All Wiechman wanted to know was what was in the little package. Why did Booth stop by to give Mrs. Surratt a package? The widow did not say, and yet, if she had wanted to hide the package, she could have put it in one of her big carryall bags. Then too, if the contents were of any consequence, why was it that, a few moments after Booth had gone, Mrs. Surratt almost forgot the package and asked Wiechman to "Wait, Mr. Wiechman. I must get those things of Booth's."

They were on the far side of the Navy Yard Bridge when Mrs. Surratt asked Wiechman if he knew that her son John was away—far away—on a trip. Wiechman said that he had not seen John around the house in some time and assumed that he had left Washington City. The mother said that he would not be back for a while. Wiechman, without saying so, assumed that she meant he was in Canada because he had often heard John speak of Canada in conjunction with the Confederate Purchasing Commission.

In Washington, Mr. Booth walked five blocks to Herndon House and upstairs to the corner room where Lewis Paine lay open-eyed on the bed, hands clasped behind head. Booth conferred with him; told him the plan for tonight, and advised him to check out of the room. A horse would be in the little stable behind Ford's Theatre. They set a time for meeting, and Booth left.

An actor came out front to the box office and told Harry Ford that the rehearsal was over, and he could start decorating the State Box. Mr. Ford asked someone to tell the ticket seller, Tom Raybold, to do it but word came back that Mr. Raybold was ill and might be in later. Ford dropped his work and went backstage and ordered Ned Spangler and Jake Ritterspaugh, a new employee, to help him. Unasked, Joseph Burrough came along too. Mr. Burrough was called Johnny Peanut because, when the theater first opened in 1863, that's what he used to sell out front.

The partition was removed and again Boxes 7 and 8 were one. A cane chair was placed outside the little white door, facing the dress circle, for the President's guard. Joe Simms, a colored boy, was sent upstairs to the Ford apartment to get the red upholstered rocker reserved for the President. He carried it through the alley upside down, on his head. Below the stage, in the subterranean passage, were two sofas which were also brought up and dusted and put in the box. Three good chairs were placed along the back of the double box.

The sofas were set in the front of the boxes, facing the stage. These were for the ladies. An upholstered chair was set between them, for one of the gentlemen. The President's rocker was placed at the very rear of the second box—number 7, in a corner where the President would be hidden from the audience by a heavy contoured drape.

Ford got onstage and climbed a stage ladder to set the flags in place. He draped big American banners across the billowing façades of both boxes. He had a small jack staff and a socket between the boxes and here he set a Treasury Department regimental flag. The space over it was bare, and Captain Jones had given him a small steel engraving of President George Washington. He called for hammer and nail and hung this picture over the Treasury flag.

Then he stepped down, pulled the ladder away, and surveyed his work. He asked for comment from Spangler and the others and they said that it looked impressive. Mr. Ford had more bunting to hang, but, after reflection, he decided not to overdo the decorations. He wished that he had a lithograph of General Grant, or perhaps Grant's first regimental flag, but he hadn't. It would have to do, just the way it was.

Mr. Ford went back into the box and checked over every item of furniture, to be sure that they were clean and placed exactly right. Then he opened the door to 7, and closed it. He opened the door to 8, and closed it. Everything was in order. It didn't occur to him that neither door locked.

Ford's Theatre was ready.

3 p.m.

Lewis Paine walked into the sitting room at Herndon House. Mrs. Murray, the wife of the proprietor, was sitting at the desk. The gladiator said that he was leaving for Baltimore. He never wasted time on pleasantries or politeness, and expected neither in return. She asked if he wanted his bill. He said that was why he was here.

Mrs. Murray made it out, and Paine paid it. He turned to leave and the lady said that it was too early for the regular dinner, only three o'clock, but that she could have some cold beef and potatoes sent up to the dining room. Paine said all right.

He ate and left.

A man walked into the President's office, at the White House, with a ready-made smile. He looked behind the desk and saw no one, and the smile died. This was Chaplain Edward D. Neill, of the First Minnesota Infantry. In the doorway behind him stood a colonel from Vermont. The chaplain was now a clerk at the White House, and had used his favorable position to have Lincoln brevet the colonel a Brigadier General.

Now, finding the office empty, he took the further liberty of examining the papers on the President's desk, trying to find the signed commission. He was shuffling papers as quietly as possible when the President, as quietly as possible, approached the desk. Chaplain Neill looked up, startled. Mr. Lincoln was chewing on an apple.

Neill explained that he was looking for the signed commission, that he was sure that the President wouldn't mind and, as he was talking, Mr. Lincoln pulled the bell cord.

"For whom are you ringing?" the chaplain said nervously.

There was a hint of humor in Lincoln's eye as he grabbed the reverend by the lapel, leaned close, and, in a stage whisper, said: "Andrew Johnson!"

The chaplain said: "I will come in again," and left.

The thing that irked Lincoln was Johnson's acting like a one-man proletariat. Johnson never tired of telling of his humble beginnings, as though it were something exclusive to him. He acted like a swamp snob and, to Lincoln, the original rail splitter, the penniless log-cabin rustic, it was anathema.

The Vice President had sought interviews with the President and he worried when, on pretexts, they were denied. When Congress adjourned, he had stayed on in Washington City, hoping to heal whatever sore was on Lincoln's mind. Now he had been sent for, and he was early and had paced the grounds outside waiting for the Cabinet meeting to be over.

Johnson felt relieved when the Chief Executive grabbed his hand and shook it vigorously and called him "Andy." The two sat on opposite sides of the desk and talked for twenty minutes. Neither told what this conference was about, but the President was not a man to mention Inauguration Day. When the appointment was made, Lincoln had told his secretary, John Nicolay, that reconstruction was a brand-new problem and that the Vice President should be made acquainted with it and should understand the wishes of the President.

The probability is that Lincoln timed the first interview immediately after the first peacetime Cabinet session so that he could not only tell Johnson what was in his mind about reconstruction, but also what the Cabinet had agreed upon. When the Vice President left, he looked less tense than when he entered the White House.

At the outer gate, a big-handed colored woman, sick with hunger, staggered up to the sentry box. A soldier stepped out, swung his rifle diagonally across his body, and said solemnly: "Business with the President?" The other guards laughed.

"Before God," the woman whispered. "Yes."

"Let her pass," one of the guards said, still laughing. "They'll stop her further on."

The woman staggered through, and across the drive and on up to the White House porch and the big doors. There, a second soldier barred her.

"No further," he said. "Against orders."

With what seemed to be her final energy, the woman darted under his arm and ran straight through the long corridor and down the length of the carpet and upstairs, her cotton skirt billowing in back, her face a study in dark agony. The presidential office guard stood before the door with his rifle in both hands.

Breathing hard, and crying at the same time, she begged: "For God's sake, please let me see Mr. Lincoln."

"Madam," the guard said. "The President is busy. He cannot see you."

She may have screamed, or sobbed, because the noise she made caused the door to open and, through the iridescence of her own tears, she saw the wavering figure of the President of the United States.

He looked down at her, and he was smiling. In his deep tones, he said: "There is time for all who need me. Let the good woman come in."

She was Mrs. Nancy Bushrod and, when he had seated her beside his desk, it took time for her to compose herself. She told him her name and how many babies she had and that her husband's name was Tom Bushrod and that both of them had been slaves on the Harwood plantation outside of Richmond. When they had heard about the Emancipation Proclamation, they heard it said that it meant that they were free, and they had run away and come straight to Washington.

In thankfulness to Mr. Lincoln, Tom had forthwith enlisted in the Army of the Potomac, leaving Nancy in a little shack with twin boys and a baby girl. His pay kept coming regularly every month, then it stopped. Nancy had walked the streets of Washington City, looking for work. No one had any washing or sewing or cleaning to be done. All the colored people in the world seemed, to Nancy, to be looking for work in Washington City.

She started to cry again. Would the President please help her about Tom's pay? She would not have come, but there was no other way.

"You are entitled to your husband's pay," Mr. Lincoln said. "Come this time tomorrow and the papers will be signed and ready for you."

As Mrs. Bushrod told about it later, she said: "I couldn't open my mouth to tell him that I was going to remember him forever and I couldn't see because the tears were falling."

When he had escorted her to the office door, he said: "My good woman" (in the tone of the stern lecturer), "perhaps you will see many a day when all the food in the house is a single loaf of bread. Even so, give every child a slice and send your children to school."

Then, as Nancy Bushrod looked back at him, she said he bowed "like I was a natural-born lady."

A little more than a hundred yards to the west, Mrs. Stanton, out shopping in her carriage, stopped at the War Department and asked her husband *how* she should decline the invitation of the Lincolns to the theater. The secretary must have assumed that the matter had been taken care of, because he looked, for a moment, as helpless as most husbands in similar situations and said: "Why, just send regrets."

No snub was intended in the regrets. Stanton had an aversion to the theater. As a serious student of the Bible, he was sure that theaters and sin were kin. He had felt, all along, that Lincoln did not expect him to accept the invitations, but had sent them along for form's sake. In all the time that Stanton had been in Washington, he had been to a theater once, and that occurred one night when he tiptoed into the State Box at Grover's Theatre with a message for Lincoln and, delivering it and getting a reply, tiptoed out.

Then too, he knew that Mrs. Stanton felt tense in the presence of the First Lady, and this alone inclined him to turn down all but invitations to state functions, where the presence of the Secretary of War and his lady were required. Thirdly, Stanton's repeated admonitions to the President to keep away from the theater, and from all public appearances, would have little validity if the man with the perfumed whiskers did not practice what he preached.

Grant stopped in to say good-by to his boss, and to leave a temporary forwarding address at Burlington, N.J. Stanton and the general talked for a few minutes about army business, and then shook hands and parted.

The general got back to the Willard at about the time that Wilkes Booth was entering Kirkwood House, two blocks east. Booth wanted to discuss final plans for the assassination of Johnson with George Atzerodt. The clerk told Mr. Booth that Atzerodt was out.

The actor had a drink at the bar, and got an idea. He reasoned that if Lincoln was killed tonight, Andrew Johnson would become President. Johnson was to be shot by Atzerodt, but Booth had little faith that this would be accomplished. If Johnson lived, the affairs of the Union would be carried on with little impediment except grief for Lincoln. Booth had to hurt Johnson's chances of becoming President, and he seized upon a novel idea.

He asked the desk clerk if the Vice President was at home. The clerk said no. Booth asked for a blank card. On it he wrote, "Don't wish to disturb you. Are you at home? J. Wilkes Booth." The actor had never met Andrew Johnson but, after the assassination of Lincoln, the Vice President would have a difficult time explaining the visiting card from the arch-conspirator. With passions high, it might look as though Johnson was the prime mover in a plot to make himself President of the United States.

Before leaving, Wilkes Booth wrote a note to Atzerodt, and shoved it under the carriage maker's door. He walked out jauntily, certain that the government would be so hamstrung with suspicion tonight that Johnson might, instead of being sworn in as President, be thrown into the Federal Penitentiary at the foot of Greenleaf's Point.

History often hinges on small accidents. A short time after the actor left, Colonel Browning came into the lobby of Kirkwood House. The colonel was secretary to the Vice President. He stopped at the desk for his mail. The desk clerk had standing orders to put all mail for Johnson into Colonel Browning's box because the colonel did all the sorting and reading for the unhappy tailor. When Browning got to his room, and began to read the missives, he came to the card that Booth had left, and at once assumed that the actor had left it for *him*. Years before, he had witnessed a performance of Booth's in the South, and had been introduced to the actor. Browning was pleased that Wilkes Booth remembered him, and made a mental note to find out if he was starring in a play in Washington. If so, Browning wanted to return the compliment.

4 p.m.

The city looked as though all of the people, in concert, had agreed to stay home until this hour. The sun was hesitant, but the air was warm and the wooden walks were crowded with women, many walking in pairs, their full skirts bobbing forward and back with each step. Buggies, surreys and gigs paraded the Avenue, and single-mount riders posted at slow trots and took in the sights. Bars were again crowded, and the underside of bristling mustaches sparkled with foam. The restaurants did a good business, for this was the supper hour. Later, the oyster houses along Maine Avenue would be catering to the trade because it was the custom, after a big dinner, to adjourn to an oyster house for seafood and drinks and leisurely conversation.

The President had finished a day's work. He had no more afternoon appointments and he got up slowly from his desk, put his fragile glasses in a case, and went to the little closet on the north side of the office to wash. He removed his black coat, pulled his cuffs loose from the links, and washed his hands. His face, if he bothered to look at it, was tired. He wet a brush and stroked it through his black scraggly hair.

Mrs. Lincoln wanted to go for a carriage drive. Well, maybe it was a good idea. He was happy and fatigued, and a relaxing drive could erase fatigue without mitigating the warm feeling that a great deal had been accomplished on this day. He was readjusting his cuffs when the Assistant Secretary of War, Charles A. Dana, walked into the office. He had been sent, he said, on a special errand by Secretary Stanton.

The President replaced his towel on a hook.

"What for?" Lincoln said.

Dana, a balding dandy whose long stiff beard jerked as he spoke, said: "Well, sir. Here is the Provost Marshal of Portland, who reports that Jacob Thompson is to be in that town tonight, and inquires what orders we have to give."

Thompson, an ex-United States Senator from the state of Mississippi, had spent most of the war in Canada as the Confederate States of America representative. He had been a thorn in the side of the U.S. Government, and now he had crossed the border into Portland, Maine. He planned to take a steamer to Liverpool.

"What does Stanton say?"

"Arrest him."

"Well," the President said, closing the closet door. "No. I rather think not. When you have an elephant by the hind leg, and he's trying to run away, it is best to let him run."

Dana thanked Mr. Lincoln, and hurried back to the War Department with the order. Stanton listened impatiently. For four years, he had waited for the day when he could lay his hands on the leaders of the rebellion and now the President wanted them to get away.

"Don't send any reply," he said. And none was sent. Thus, if the Provost Marshal at Portland, in the absence of orders, took it upon himself to arrest Jacob Thompson, Stanton would have the rebel exactly where he wanted him and, at the same time, could not be accused of disobeying orders.

The marshal must have had some experience in politics. He permitted Thompson to get away.

Wilkes Booth had done a lot of walking this day. Now he would ride. He walked down the Avenue, and across the Mall to Pumphrey's Stable. His mare was ready. She was muscular and excitable. As the groom cinched the saddle under her belly, she flattened her ears and tried to bite him. When the stirrup length had been adjusted, the groom slapped her flank and she jumped. The actor walked around her, an imposing figure himself in black hat and black, smartly tailored coat and black, tight-legged trousers. The tan boots were brightly polished and the spurs gleamed like gems. He examined the fit of the bridle and ran fingers under the edges of the light blanket beneath the English saddle. He walked her in a circle, watching her feet.

She was a good mare, solid, with a deep chest and nimble legs. He would have preferred the horse he had been riding for weeks, but the mare looked as though she had stamina, and, between Ford's Theatre and the far side of the Navy Yard Bridge, she was going to need every ounce of it.

He mounted and pulled on one rein. She turned around in her own length. Booth kicked his feet out of the stirrups and asked the stableboy to shorten both one more notch. He sat easily and well and the mare and her burden moved out of the stable lightly in an odor of leather and brandy.

The actor walked her awhile, up Sixth Street to Pennsylvania Avenue. Then, swinging left, he gave her a touch of the spurs and she quivered and went into a run. He pulled her down to a trot, pleased that she did not have to be urged to move fast. Booth was in the stream of carriages and horsemen moving up the Avenue. On the other side of the street, two more regiments of Union soldiers were coming into the city, moving to bivouac areas.

Where E Street meets the Avenue, Booth turned his mare at Grover's Theatre and cinched her to a hitching post. He went into the manager's office looking for Dwight Hess, but it was empty. He went upstairs to Deery's tavern and asked for a bottle of brandy and some water. Deery said that he had Booth's box seat for tonight and Wilkes told him that he would pick up the tickets later. He drank deeply and looked around at the few men who played billiards. As he left, he spun a cartwheel coin on the bar.

Downstairs, the manager's office was still empty. John Wilkes Booth was ready to write another letter. He sat at the desk and withdrew paper and an envelope from the pigeon-holes. The earlier letter left with his sister Asia would explain his motives, which had not changed, but that letter was designed to explain the capture of the President and the heads of the government. He needed a new one to explain murder.

What he thought about, no one knows. Across the envelope, he scribbled "Editor, National Intelligencer." Now, he had to think more seriously in terms of his own death. He was well known to many thousands of people and there would be no difficulty in establishing his identity. If he should be mangled beyond recognition, a tattoo across the front of his left wrist said "J.W.B."

He wanted no confusion about the perpetrator of this deed. The two most important factors in it—after the commission of the deed itself—were to let the world know that it was John Wilkes Booth who did it, and to let the world know why he did it.

The words, once begun, came swiftly to the skating pen. He had devoted time and effort, he wrote, to effect the capture of this man; now he found that he had to change his plans to give the South one more chance. He expected criticism of his act, but someday, when sectional anger had cooled, time would justify him.

Then Booth did a mean thing; an ignoble thing. Instead of signing it with his name alone, he decided to commit his friends, his fellow conspirators, to his deed. He wrote:

"J. W. Booth—Paine—Atzerodt—Herold."

He knew that this put the noose around their necks, and he felt no compunction. If the others withdrew from the plan, from this moment onward, they had been named as parties to it and, if Booth killed the President, the others would hang for it. Had he been able to add to the list, in justice, the names of Arnold, O'Laughlin, John Surratt, Mrs. Surratt, there can be no doubt that he would have been no more solicitous about their welfare than he was of the others.

Booth sealed the letter, put it in his pocket, and went outside and started off down E Street. The next stop was to be Ford's Theatre.

The President sent word to Mrs. Lincoln that he would be at the War Department for a few minutes and, when he returned, would be ready to take a drive with her. She wanted to know if he would like to invite some friends along. He said no, "Just ourselves."

Crook walked with him out to Pennsylvania Avenue and to the left, toward the War Department. They approached a group of celebrants on the walk, and it was obvious that these men were in the violent, argumentative stage of drunkenness. Crook had to jump ahead, to clear walking room for the President. After the two men had passed by, Mr. Lincoln said:

"Crook, do you know, I believe there are men who want to take my life." This surprised the guard, because it was the first time that the President had initiated such a topic. "And I have no doubt they will do it," he added.

"Why do you think so, Mr. President?"

"Other men have been assassinated. . . ." and the voice trailed off.

"I hope you are mistaken, Mr. President."

"I have perfect confidence in those who are around me," the President said firmly, no longer in reverie. "In every one of you men. I know no one could do it and escape alive. But if it is to be done, it is impossible to prevent it."

At the War Department, the President greeted Mr. Stanton with an expectant smile and asked if there was any news from General Sherman. Stanton shook his head slowly. He too was waiting, with poor patience, for the last of the great good news. It hadn't come.

"I am looking for someone to go to the theater with me tonight," Mr. Lincoln said. "Grant says that he cannot attend, and neither can you. May I have your man Eckert?"

He referred to Major Thomas T. Eckert, chief of the War Department Telegraph Office, a man who was big all over—not just in tallness—and who had an outsized chest and arms and even ears. He looked like a cartoon of an off-duty policeman. Eckert, emotionally, was a man who was a completely subordinate subordinate. Whenever, in the normal course of events, he was given an option of action, his eyes turned at once to his boss for a sign.

Stanton said that he was sorry, but that he had important work for Eckert that night. The President rubbed his fingers through his chin whiskers and said that he thought highly of Eckert as a bodyguard because once, when the major complained about the quality of a shipment of iron pokers, Lincoln had seen him take them, one after the other, and break them over his arm.

Mr. Stanton said that Eckert's work could not be put off. Besides, the nation was in a state of turbulence at the end of a long and bitter war and the President would do well not to be seen in public at a time like this. The War Secretary counseled again against attendance at the theater.

"Very well," said Lincoln. "I shall take someone else, but I should have preferred to take Major Eckert because I have seen him break a poker over his arm."

When they left the War Department, Crook walked beside the President again en route back to the White House. The guard noted that the drunks had gone. Mr. Lincoln got to talking about the theater party. He knew that Crook had heard what the Secretary of War had said.

"It has been advertised that we will be there and I cannot disappoint the people. Otherwise, I would not go." They walked on. "I do not want to go."

Back at the Executive Mansion, Congressman Samuel Shellabarger of Ohio was waiting. He wanted one of his friends appointed to the staff of the United States Army. Lincoln said he was reminded of when he was a young fellow. A lady in the neighborhood made shirts. An Irishman asked her to make one for him. She made it, laundered it, and gave it to him. When he put it on, he found it was starched all the way around instead of only in the collar, and he sent it back saying that he didn't want a shirt that was all collar. "The trouble with you, Shellabarger, is that you want the army all staff and no army."

Young David Herold was in a hurry. He had expected to find Booth at Taltavul's or at Ford's Theatre and, missing him at both places, he dog-trotted down to Naylor's Stable on E Street. This was almost directly opposite Grover's Theatre. Wilkes had just left there and was now en route to Ford's Theatre.

At Naylor's David asked for a horse he had reserved earlier. The stable foreman, John Fletcher, was an Irishman who suspected that everyone thought he was "green" and was trying to "do" him.

"Until eight o'clock only," said Fletcher.

"How much will it cost?" said Herold.

"Five dollars for the evening."

"I heard it was four."

"Five."

"All right. Give me the roan mare."

"I have a good mare for you."

"No. I want the roan. The one I had before."

Fletcher brought the roan out and gentled her to the

front of the stable. He got a saddle, but David didn't like it. Fletcher got another one, but the boy didn't like the stirrups. He wanted a flat English saddle and open metal stirrups. Also, he wanted a double-reined bridle. When he mounted the mare, he asked how late he could stay out.

"No later than eight," John Fletcher said. "Nine at the most."

Fletcher stood in the doorway and tamped tobacco into his pipe. He didn't like this boy. He didn't trust any of them. How many times even the best of them had stolen good animals on him, even when he had a deposit. Or, if they kept the animal after hours, they would drop it off at another stable and, in the morning, Fletcher had to make a tour of dozens of stables around the city to find the missing animal. Well, he'd give this boy until 9 P.M. on the dot, and then he'd go out looking for him.

Booth arrived in front of Ford's Theatre and saw Maddox, the actor. Sitting his mare, Booth talked while Maddox stroked the horse's mane. Suddenly, Booth stopped the conversation and said: "See what a nice horse I have got? Now watch. She can run just like a cat." He dug his spurs into the roan and, in a flash, was racing down toward the Avenue.

5 p.m.

The President of the United States came out on the White House porch with Mrs. Lincoln. He studied the sky and buttoned his coat. They were using the barouche. The coachman helped Mrs. Lincoln into the carriage and, as the President followed, she said: "Would you like someone to come with us?" She had asked that before.

He got up in the carriage without assistance, settled himself at her side and tucked a blanket around both of them.

"No," he said, with a twinkle of gallantry, "I prefer to ride by ourselves today." He nodded to Francis Burns, the coachman, and the sparkling carriage started out of the gravel driveway. Two cavalrymen fell in behind the coach, but the President did not look back.

On this drive, the President was in rare humor and passersby heard Mrs. Lincoln's laughter peal from the coach. It rang out wholeheartedly and Mr. Burns, up on the front seat, started grinning without knowing the joke. They went along G Street at a spanking trot, the President raising his silk hat when groups of citizens hailed him from the walks, and the carriage turned down New Jersey Avenue, still moving at a smart clip.

"Dear husband," Mrs. Lincoln said. "You almost startle me by your great cheerfulness."

"And well may I feel so," he said, becoming serious at once. "Mother, I consider that this day the war has come to a close." He patted her hand, as though he hoped to infuse her with what he was going to say. "We must both be cheerful in the future. Between the war, and the loss of our darling Willie, we have both been very miserable."

Mrs. Lincoln stopped laughing. The death of her Willie was a sore deeper to the bone than the war had been. For a while, both were silent. The matched blacks trotted as though they would never tire and, as they passed the Capitol, both saw the marching troops, the cadres of dejected prisoners, the end of something they had lived with for a long time. The city was now relaxed. It was gay and careless and silly and in fettle. All of it was blessed with contagion.

The President, who rarely spoke of his own future, told Mrs. Lincoln that he wanted to get on with reconstruction in the South, complete his term of office, and then perhaps take a trip to Europe with his family. He would like that, he said. Then he would return to Springfield, Illinois, and perhaps resume law practice. He was happy and murmurous as he talked, half to himself, half to her. It would be nice, he thought, if someday they could buy a prairie farm along the Sangamon.

The translucent quality of his happiness was such that Mrs. Lincoln began to feel disturbed. He seemed to read her thoughts as the carriage neared the Navy Yard.

"I never felt so happy in my life," he said. It was as simple and unequivocal a statement as he had ever made, and he was noted for them.

And she, a hysteric tortured by fears, said: "Don't you remember feeling just so before our little boy died?"

At the Navy Yard, the President got out, stretched his legs, and was induced to walk the deck of the monitor *Montauk*. Then he stepped back into the carriage, nodded to Burns, and they were off for the return trip.

Booth walked his mare up the Avenue. He waved to friends, raised his hat now and then, and continued on to Fourteenth Street where he saw the actor John Matthews. This is the man who refused to become a part of the scheme, and of whom Wilkes said: "He is a coward and not fit to live." Now, he pulled rein and smiled and reached down to shake Matthews's hand.

The star dismounted and slapped dust from his trousers. They talked of plays and bookings. Matthews was flattered to be seen on terms of intimacy with Booth, and he was determined not to bring up the matter of the ridiculous plot that Wilkes had told him about.

Booth had a favor to ask—a small thing. Would his friend John deliver a letter for him tomorrow? He withdrew it from his pocket. It was just a note to the editor of the *National Intelligencer* and he would have done it himself except that he expected to be out of town tomorrow. Would Matthews do it? Certainly. Matthews would be glad to do it—could do it now if it would help.

Oh, no, said Booth. Not now. I would do it myself if it could be delivered now. The letter contained a matter of news for the *National Intelligencer* and it must be delivered tomorrow, preferably just before noon. Matthews took the letter and told his friend to consider it done.

"What's so important about it?" Matthews said.

Booth was looking across Pennsylvania Avenue at a new group of marchers. They were bedraggled and dejected.

"Who are those men?" he said.

Matthews squinted.

"They look like officers of Lee's army."

Booth mounted his mare and looked despairing as he swung away.

"Good God!" he said. "Matthews, I have no country left!"

The marching men looked sullen and some thrust their hands into their tunics as they shuffled along. A few limped.

John Matthews put the letter in his pocket.

Booth raced up to Fifteenth Street, turned and started back slowly, just as a carriage pulled away from Willard's. Two cavalrymen rode behind the carriage. The actor knew that none but important government personages used outriders, so he spurred his horse and passed the carriage, which was headed down the Avenue toward the Capitol. As he flew by, he looked in the carriage and saw two women. One was Julia Dent Grant. A woman friend was seeing her off at the station. On the box, up front, was the coach driver and Lieutenant General Ulysses Simpson Grant.

The actor reined his horse, some distance ahead, and then turned and came back toward the carriage at a walk. He stared hard at Grant, and he looked into the carriage at the ladies with such intent that both of them remembered the strange, wild-eyed rider later. He nodded to the cavalrymen and went back to Willard's corner. He said to a man:

"Wasn't that Grant?"

The man nodded.

"I thought he was going to Ford's tonight, with Lincoln."

The man shrugged. "Somebody said he's going to Jersey."

The only railroad running between Washington and Burlington, N.J., was the Camden & Amboy. There were four northbound trains daily. The one chosen by the general was the slowest.

The schedule:

Lv. Washington City	6 P.M.
Arr. Baltimore	7:25 P.M.
Arr. Philadelphia	12 midnight
Change trains	
Lv. Philadelphia	6 A.M.
Arr. Burlington	7 A.M.

Total time: thirteen hours. If the Grants had remained at the Willard overnight, they could have boarded the 7:30 A.M. express and would have been in Burlington, N.J., at 2:58 P.M. after seven and a half hours of travel and a night's rest.

The Surratt carriage arrived at John Lloyd's tavern and, as Mrs. Surratt alighted, Wiechman said that he would like to run the horse up and down the road. The widow assented and took the small package with her. Inside, she met Mrs. Offutt and she asked for John Lloyd and learned that he was appearing in court at Marlboro that day.

She asked about Mr. Nothey and was told that no one knew his whereabouts but that he had been around telling neighbors that he had tried to settle the widow's claim of $479, but that she didn't want to reach an agreement. This confirmed what she had learned by mail and it outraged her because, at the boardinghouse, she had been existing on a day-to-day basis and, in Maryland, Mr. Calvert had won two judgments against her because she couldn't pay money she owed.

Wiechman came in and Mrs. Surratt asked him to write a letter for her. It was posted at once:

Surrattsville, Maryland, April 14, 1865
Mr. John Nothey
Sir:
I have this day received a letter from Mr. Calvert intimating that either you or your friends have represented to him that I am not willing to settle with you for the land. You know that I am ready and have been waiting for the last two years, and now if you do not come within the next ten days, I will settle with Mr. Calvert and bring suit against you immediately. Mr. Calvert will give you a deed on receiving payment.
M. E. Surratt
Administratrix of J. H. Surratt.

According to the trial records, Lloyd and Wiechman testified to the following:

At 5:30, John Lloyd, home from court and a card game, drove up the farm lane beside the tavern and drew to a stop beside the kitchen wood pile. He got out, staggering, and managed to lift a bag of oysters. Mrs. Surratt saw him, and came out by the kitchen entrance and met him halfway down the flagstone walk. She was smiling.

"Talk about the devil," she said, "and his imps will appear."

Lloyd glared and tried to focus on her. "I was not aware that I am a devil," he growled.

"Well," she said, "Mr. Lloyd, I want you to have those shooting irons* and some whiskey ready. There will be parties here tonight who will call for them." She handed the small, paper-wrapped package to him. "Hide this, for tonight."

He had trouble holding it, and the oysters too. He went inside, Mrs. Surratt holding the door for him, and when he got to the sitting room Lloyd fell on a couch. He felt ill. Nausea hit him in waves. After a while, he raised himself on an elbow, panting.

The widow came in, looked, and shook her head sadly. She said that a spring had broken on her buggy and could John do something about it. Lloyd tried to say that he would tie it with rope but the best he could do was to mumble "Rope. Rope."

Somehow, he managed to stand, and somehow, he summoned the required amount of coordination to walk out front and tie the buggy spring so that it would hold together. He stood on the porch to wave farewell as Wiechman and Mrs. Surratt started back to Washington City.

Lloyd, head and hands shaking, staggered back into the bar and asked bartender Joe Nott for a glass of whiskey. He got it down and kept it down and then he remembered the little package in his hand. He hid it upstairs with the

---

* Again, this is Lloyd's testimony. Largely because of it, Mrs. Surratt was hanged.

other stuff, after peeling the wrapping paper off and taking a look.

Field glasses.

In the War Department, Mr. Stanton buttoned his coat and went inside to the telegraph office.

"I have changed my mind about tonight, Eckert," he said. "I will not return."

The telegraph chief always stood in the presence of the Secretary of War. Mr. Stanton did not tell him what the important night duty was to have been.

"Yes, Mr. Stanton," said Eckert.

"Good night, Eckert."

"Good night, Mr. Stanton."

The secretary did not have to say anything about night orders for the telegraph office. The men on duty knew those orders. No matter how late the hour, the secretary was to be awakened at home and acquainted with any important news. If there were any messages addressed to President Lincoln, the same rule applied.

On a side street, Booth saw a familiar figure on foot. He drew up alongside George Atzerodt and dismounted and held a whispered conference. He told George that the "elimination" of Lincoln was easier now, because Grant had just left town. It was important, he said, that Atzerodt time his attack on the Vice President for 10:15 P.M. or as near to that minute as possible, because the other attacks would be made at the same time, and if all hands finished at the same time, they would meet on the far side of the Navy Yard Bridge at almost the same time.

Atzerodt, drunk and brazen, said that he "enlisted" for capture, not murder. Booth sneered at the carriage maker and called him a coward. George started to whimper and said that he had spent a day investigating the Vice President and he had learned that Mr. Johnson was a brave man. Booth told him that he had gone too far already, and might as well go the rest of the way.

The little man agreed. "I am in trouble," he said, "and I will never be shut of it." Booth slapped him on the back and wished him luck. The actor got on his mare and rode off. Atzerodt watched him go, probably wondering if he would ever see him again, knowing that he wouldn't. At this moment, he knew lots of things. He knew that he would go back to Kirkwood House, deceiving himself as to his intentions, and that he would drink and debate whether he had the nerve to kill Johnson and then, finding that he did not, he would drink some more and curse himself for being a coward. He remembered what the men in Port Tobacco had said of him: "George is a man who can be insulted without taking offense."

He walked to a stable where he had a horse. He had rented the horse at one stable, and left it at another. He considered this necessary to a complicated plot.

Atzerodt drove out of the stable, and up past Ford's Theatre, which he studied as he went by, and shook his head. He turned and passed the Patent Office and, at that point, tossed away a bowie knife in a sheath. He decided then to ride aimlessly for a while, pausing at different saloons and only having one or two drinks in each one.

The only use George Atzerodt was to anyone now was that he was a man who knew a secret.

6 p.m.

The day was dying the way it was born—gray. The warm breeze spent itself and there was a stillness and a coolness and flags hung lifeless and cloaked the staffs with color. The smoke from an outbound train hung like a crayon apostrophe behind the Capitol and farmers came in from outlying counties to sell feed to the stables.

Many of the principals of this day were, at this time, on the road. Mrs. Surratt and Wiechman were bound for Washington City. The President and his lady were being driven back to the White House by a different route than they had taken to the Navy Yard. Atzerodt, on a horse, was in search of friendly faces. David Herold had left Naylor's Stable and was looking for Booth. Paine was wandering, waiting for the proper time to meet his idol and pick up his horse. Stanton was homeward bound, a few blocks from his office. So too was Major Thomas Eckert. John Wilkes Booth dismounted at F Street between Tenth and Ninth, opened an old billboard gate leading to the alley behind Ford's Theatre, led his horse inside, and slammed the gate, which hooked on a latch.

He rode slowly down the alley, between the shanties of the Negroes, carefully skirting the little groups of children playing, passed the backs of the Ninth Street boardinghouses with their clothespoles and outdoor lavatories, and dismounted at the back door of the theater.

He shouted "Ned!" and, in a moment, Spangler came out and so did James Maddox. Booth asked his friend to stable his mare and he asked for a strong halter because she was known to shred them and run away. Spangler said he had a good one in the property room and yelled to Jacob Ritterspaugh to get it for him.

When the mare had been given water and feed, Booth invited the stagehands next door to Taltavul's for a drink. At this hour, only Spangler, Ritterspaugh and Maddox were in the theater, with the exception of a ticket seller out front. The conspirator invited everyone except the ticket man to join him in a drink.

He led them backstage, then down into the subterranean passage single file, then out through the south alley to the tavern. There he bought drinks and learned that a boy had been in looking for him—Herold. Booth had the respect of the stagehands because he was friendly without being condescending. He joked with them and, in conversation, asked if they had to be onstage for any particular work right now, and they said no, that this was dead time, that the scenery for tonight had already been set up and they were killing time.

Booth said that he had an errand to do, but, before he left, he bought a bottle of whiskey for the men and advised them to "drink up." He left the group, saying that he might see them tonight, and went back through the alley and through the underground passage and up onto the stage. He picked up a 11/2-by-3-inch pine board which had held a music stand. A single brace of gas lamps was burning over the stage and, in their feeble light, he could see the flag-draped box.

He hopped offstage onto the orchestra floor and walked toward the back of the theater and on up the stairs to the dress circle. He moved along the aisle to the south and down to where a cane chair sat before the white door leading to the State Box. The actor paused, looked around, and saw no one.

Booth tried the door and it opened easily. He went inside, and closed it behind him. He struck some matches and tried to brace the pine board between the inside of the white door and the rear wall of Box 7, which jutted into the corridor. The board was about half an inch too long. He had a penknife, but it would take an interminable time to whittle the board and time now was of the essence. Again he braced the board against the door—just above the knob—and tried to fit it against the wall. Where it jammed, he gouged plaster from the wall, holding a kerchief just below the spot and catching the fragments. In a few minutes, he had a niche into which the board fitted well.

Tonight, if anyone tried to follow him from the dress circle, they would find that the harder they pushed on the door, the more firmly wedged the board would become. He expected that, if he had to stab the guard outside the white door, there might be an outcry and people would try to follow. The board was important. He removed it and set it in a dark corner near the door.

Next he looked into the box. The partition was gone, the sofas and chairs had been arranged, and here, in the rear of the box, was President Lincoln's rocker. Both box doors were tried, and both worked easily. As he had known all along, the locks were broken.

Now he stood between the President's rocker and Mrs. Lincoln's chair and studied the jump to the stage. It wasn't much. If he stood on the ledge and jumped, it would be two feet higher, but if he swung himself over the ledge, hung on by his hands, and dropped, it would be shorter and safer.

John Wilkes Booth went out into the little corridor, with the door open, and crouched and studied the position where the President would be sitting. Then he closed the door and lit some matches. He withdrew a spiral-fluted gimlet with a wooden handle from his jacket pocket. The door had two panels, a top one and a bottom. At its thickest, the door was three-eighths of an inch deep. The panels were recessed and considerably thinner. He set his gimlet against the lower right-hand corner of the upper panel and leaned against it.

He turned the handle and shavings began to fall off. In a moment, he had a small hole drilled through. He pressed his eye against it, with the door closed, and he had a hazy view of the upper part of the rocker. From his pocket he took a penknife, and began to gouge and ream the little hole and to peel the shavings from it. Again he stooped and looked. This time he had a good view of the spot where the President's shoulders and head ought to be.

Booth struck some more matches, and scooped up the shavings. He went back to the entrance to the corridor and, under matches, looked for grains of plaster. He scooped up the dust, dropped it into his pocket, and stepped outside the white door into the dress circle. Carefully, he studied the stage, the orchestra, and the dress circle. No one was in sight and he could hear no one.

He went downstairs and out the back way and got his mare. This time, instead of going up to F Street and opening the gate, he swung the other way in the T-shaped alley and went out Ninth Street, where there was no impeding gate. He went back to the National Hotel to eat and rest.

Mr. and Mrs. Lincoln were pulling into the White House driveway when the President saw two men leaving. They were two old friends from home—Dick Oglesby, new governor of Illinois, and General Isham N. Haymes—and Mr. Lincoln stood in the barouche and yelled to them.

At the door, Mrs. Lincoln left the men, and advised the President that supper would be ready in a few minutes. Lincoln said that he wouldn't be long and he took his cronies into the office. The two, seeing the one so unusually happy, fell into a mood of horseplay and, in reminding each other of old events which only they would remember, all three roared with laughter.

The President had a column clipped from a newspaper and he said that it was one of his favorites and he would read it to his friends. It was written by David R. Locke, who wrote coarse dialect under the name of Petroleum Vesuvius Nasby. Sometimes, the President would read these columns in bed, and would laugh hard and slap his thigh and, wiping tears from his eyes, would appear at the bedroom door in his night-gown looking for someone to share them with.

This one read:

"I survived the defeet uv Micklellan (who wuz, trooly, the nashen's hope and pride likewise) becoz I felt assoored that the rane of the Goriller Linkin wood be a short wun; that in a few months, at furthest, Ginral Lee wood capcher Washing-ton, depose the ape, and set up a constooshnal guvernment, based upon the great and immutable trooth that a white man is better than a nigger."

The Confederates had "consentratid" and had lost Rich-mond. "Linkin rides into Richmond! A Illinois rale splitter, a buffoon, a ape, a goriller, a smutty joker, sets hisself down in President Davis's cheer and rites dispatchis! . . . This ends the chapter. . . . The Conferasy hez at last consentratid its last concentrate. It's ded. It's gathered up its feet, sed its last words, and deceest. . . . Linkin will serve his term out—the tax on whiskey wont be repeeled—our leaders will die uv cha-grin, and delerium tremens and inability to live so long out uv offis, and the sheep will be scattered. Farewell, vane world."

Although there were barbs in the copy, aimed at him, the President almost always laughed at these stories. Once, he said: "For the genius to write these things, I would gladly give up my office." On another occasion, he said: "I am going to write Petroleum to come down here, and I intend to tell him if he will communicate his talent to me, I will swap places with him."

Word came that a cold supper was waiting and Governor Oglesby and General Haymes begged off and said that they had appointments, that they stopped in just for old times' sake, and would see the President again over the weekend.

The Lincolns, with Tad and Robert, ate together and Robert said that he had an evening out scheduled and did not know whether or not he could use the tickets to Grover's Theatre. This reminded Mrs. Lincoln that she had invited a young engaged couple, Clara Harris and Major Henry R. Rathbone, to come with them to the theater. Miss Harris was dark and lovely, a full-figured girl with rows of tiny spitcurls on her forehead. The major was tall and slender, a quiet, handsome man. It was a unique love affair. Miss Harris was the daughter of Senator Ira Harris of New York. The major was the Senator's stepson. Mrs. Lincoln said that, en route to Ford's Theatre, they would pick the couple up at the Harris home on H Street near Fourteenth. The President received the news in silence and nodded.

It was getting dark when, a hundred miles to the south, Marshal Ward Hill Lamon, the presidential worrier, drove through Richmond for the first time. The President was in his thoughts often, and there is no doubt that, as he saw the burned-out shells of once fine homes, and saw the deep strain of bitterness in the faces on the walks, he shuddered when he thought of the President walking these streets a week ago.

And if he continued to think of Lincoln, he must also have thought of the changes which had slowly come over the President, changes which old Hill saw, and which others, including Lincoln, could not see. He saw Lincoln coming to believe in his destiny as a great man, coming to believe in portents and dreams, coming to believe that it had been written in the stars scores of centuries ago that he was to be cut off at the very height of his fame and power, coming to believe that he must die at the hands of an assassin.

Hill was particularly vexed at the presidential belief in dreams as a portent of good and evil. When he told a dream in unsympathetic company, Lincoln sensed it and made fun of the dream and said that everybody knew that they had no meaning, but it was obvious to Lamon that Lincoln not only believed in them, but was preyed upon by them and worried about them.

Once the President had said that he believed that dreams were part of the "workmanship of the Almighty." If Lamon made fun of this thesis, Lincoln would turn his heavy-lidded eyes on his old friend and say: "Hill, play a little sad song for me," and Lamon would plunk it out on a banjo, the two of them listening to the cracked notes, the plaintive air.

The President said that, before he came to the White House, he was lying on a couch in Springfield and he glanced up at a mirror and saw two images of himself: one glowing bright, one ghastly in death. The meaning, he said, was decipherable: he would be healthful in his first term of office, and death would overtake him in his second. He admitted that, since that day on the couch in 1860, he had tried many times to conjure the same double image and had failed.

Dreams were in code; they were warnings, clues, waiting to be understood. He could plumb their meanings and he believed that the art of understanding dreams was shared only by common people. "The children of nature," he called them. He had a deep respect for the wisdom of the "children of nature" and believed that they were wiser, in many ways, than those with vast formal educations.

As he rode through the streets of Richmond, Ward Hill Lamon thought of the dreamer back home in Washington City. He loved Lincoln, and he knew that, if the President went outdoors on any of these nights, he would never forgive himself for not remaining at his side.

Ford's Theatre opened. The sun was setting at 6:45 and Peanut John came out front and lit the big opaque gas globe in front of the main entrance. A small, steady line of people stood before the box office. The ushers were dusting the gas globes around the walls inside the theater and Spangler, Maddox and Ritterspaugh sat out on the stone step in back, each pleasant in the glow of whiskey and each wondering if Booth would come back later and buy some more. John Matthews arrived backstage and removed his frock coat, the one with the letter in it. Harry Ford went out to the box office to remind Joseph Sessford—who was substituting for the sick Thomas Raybold—not to sell any box seats for tonight's performance.

The Night Hours

7 p.m.

Night came like a gentle sneak and the city lamplighters fought it with ladders in one hand and tapers in the other. Women about to go out for the evening studied the feeble lakes of light in the sky and decided to take good warm coats; some even carried small muffs. The people were still on the town, still celebrating, and with darkness came license. The rented carriages did a fine business, merely standing in front of a saloon until a couple came out, and those who had never hired one before hired a carriage tonight and, in the sour perfume of partly digested whiskey, men kissed their girls passionately and made promises which, in the morning, would not be remembered.

It was another gala night in a succession of them. Man had been harassed and fatigued by war for so long that he had forgotten that this was not the normal way of life, and now that he was free of anxiety, he was excessive in his desire to taste what it was like. Those who had least believed that this day would come made the most of it.

In the White House, William H. Crook stood outside the President's office with hands clasped behind his back. He was angry, but he was good at hiding his feelings. He had worked the 8 A.M. to 4 P.M. tour of duty as Lincoln's bodyguard today, and, since 4 P.M., he had been waiting for the night man to relieve him.

John F. Parker was now three hours late. Crook would not complain to Mr. Lincoln, nor to anyone else. This had happened before, and John Parker's shiftlessness as a member of the Washington City police force was well known. No one seemed to know who had selected him as a personal body-guard for the President. His record, as Crook knew, was bad, but that too seemed beside the point right now. The only thing that mattered was that John F. Parker was three hours late.

William H. Crook was still thinking about it when he saw Parker walking down the corridor. The good policeman choked down his anger and merely said: "The President is going to the theater tonight. You will go with him. Are you armed?" Parker said he was, and patted his pocket. Crook told him, tersely, the White House news of the day, and said that, after supper, he had walked Mr. Lincoln over to the War De-partment for a final evening visit, but that telegrapher Bates, and newspaperman Noah Brooks, who had been there, said that there was no news from General Sherman. Mr. Stanton, they found, had gone home. So too had Major Eckert. Grant would not be going to the theater with the President tonight; he had gone to visit his children. The President was going to take an engaged couple instead, which meant that there would be no room in the presidential carriage for Parker, and perhaps he had better leave the White House fifteen minutes ahead of time, and wait there for Lincoln.

All this he told the night man, and then, when the Pres-ident appeared at the office door for a moment, Crook said: "Good night, Mr. President." The President looked at him and said: "Good-by, Crook." On the way home, William H. Crook thought about it. The President had always said "Good night, Crook."

The night guard assumed his post outside the office. As a man, John F. Parker was somewhat less than average. He was thirty-four, sandy-haired, married, father of three small chil-dren. His outstanding virtue was that he had none. He was born in Frederick County, Virginia, had served three months with the Union forces in 1861, and lived with his family in small quarters at 750 L Street. He had been accepted as a policeman on the Washington metropolitan force early in 1861.

On October 14 of his first year, Parker went on trial before a police board for using profane language to a grocer. The superior who suspended Parker filed an additional charge of using vile and insolent language to a superior officer. Parker was found guilty, reprimanded and transferred. Later, he was charged with insulting a woman who had asked for police protection. In succession, Parker was further charged with abusing a superior officer, and with being found sleeping on a streetcar.

In April 1863, he was charged with being found drunk and disorderly in a house of prostitution, where he had appeared as a customer. The house was operated by Miss Annie Wilson and, at the trial, Miss Wilson and her coworkers appeared as character witnesses for the policeman. One girl told the police board that Parker had been staying at the house for five weeks and had not been seen drunk or disorderly in all that time. It was testified that he had been sleeping with Miss Ada Green.

When the hearings were finished, the police board found that Parker was "at a house of ill fame with no other excuse than that he was sent for by the keeper" and, in addition, found that there was "no evidence that there was any robbery there or disturbance of the peace or quiet of the neighborhood." He was acquitted.

Two weeks later, a roundsman brought Parker up on charges of not patrolling his post. He had been found sleeping on a Third Street car. Parker's defense was that he and a police partner were investigating the squawking of ducks on the streetcar and, in the midst of his investigation, he had fallen asleep. The charges were dismissed.

Ninety days later, a woman citizen charged Parker with using insulting language to her because she had protested to him against disorderly Negroes in her neighborhood. The charge was dismissed.

This is the man who drew the 4 P.M. to midnight shift as personal guard to the President on April 14, 1865. There was nothing sinister in the assignment. Since late in the previous winter, the Washington metropolitan police force had been furnishing men to the White House for this purpose. Out of a force of fifty men, four had been picked for this duty; two others had been selected as substitutes on days off. Of the forty patrolmen who might have pulled White House duty, many of the older men did not want it and, by seniority, managed to evade it. Of the younger ones, a few aspired to this assignment because it was known that the President, or Mrs. Lincoln, would keep such men from being drafted for army duty.

Three of the guards (Crook, Alexander and Parker) had asked for such letters to the draft board. It can be assumed that no police superior, including Major A. C. Richards, superintendent of the force, acquainted the Lincolns with the past records of the guards. Both the President and his wife took for granted that the character and courage of the men sent to them were beyond question.

Over on K Street, Mr. Stanton finished his dinner and said that he felt tired. Mrs. Stanton suggested that he lie down, but he said that he wanted to read the evening newspapers. He compromised by fixing pillows on a couch downstairs, and read as Mrs. Stanton sewed. Other Cabinet members were spending a quiet evening at home: Speed, Usher, Welles.

Downtown, boys ran through the streets passing out specially struck handbills.

FORD'S THEATRE
Tenth Street, Above E

Friday Evening, April 14, 1865

This Evening
The Performance will be honored
by the presence of
PRESIDENT LINCOLN

Benefit and Last Night of
MISS LAURA KEENE
in
Tom Taylor's Celebrated Eccentric Comedy,
As originally produced in America by Miss Keene,
And performed by her upwards of one thousand nights
entitled
OUR AMERICAN COUSIN

The name of General Grant, it will be noted, had been eliminated from the handbill. It is probable that the actor Matthews, returning from the Avenue, informed the Fords that Grant had been seen on his way out of town, and therefore could not be expected at the performance. Maddox, now acting as stage manager, went up to the President's Box to make a last-minute inspection, and he found everything as it should be.

Booth too was busy. He finished eating and went to his room and loaded his pistol. It was a brass derringer, a single-shot firearm with a short barrel. It would fire once, and could not be fired again without reloading. The derringer could almost be hidden in a lady's hand. It fired a good-sized lead ball, almost a half inch in circumference.

The actor loaded it carefully, and placed a percussion cap under the hammer. The derringer was ready. He patted the sheathed knife stuck in the waistband of his trousers, and assured himself that it was there. Booth had a disgust of knives, and for the type of wounds they inflicted. He kept the blade as insurance.

From his trunk he pulled a false beard, a dark mustache, a wig, an oversized plaid muffler, and a makeup pencil. When all of the accouterments of assassination had been inspected and counted, he dropped on the bed, being careful to keep his boots and spurs hanging off the edge. It is hardly possible that, with the small amount of time left for inactivity, he slept at all. Booth was sobering a little, and he probably watched the shadows on the hotel ceiling, and listened to the chatter of the merrymakers on the street below.

No one knows, or ever will know, when it first occurred to him to utter a phrase after the murder. He was not a Latin scholar, and beyond what little Latin the study of English etymology would give him, he knew none at all. He decided to use as his bid for immortality the phrase "Sic Semper Tyrannis." This was the motto of the State of Virginia: "Thus always to tyrants."

At 7:45, he got off the bed, packed his impedimenta, and, at the last moment, decided to take two big revolvers. The others might need them; he would not. Downstairs, it was said by some, he walked across the lobby, elegant and poised as always, and nodded to the desk clerk.

"Are you going to Ford's Theatre this evening?" Booth said.

"I hadn't thought of it," the clerk said.

"Ah," said Booth wagging a finger, "you should. You will see some fine rare acting."

He went out into the darkness and got his impatient mare. The night clouds were low enough to be seen in the reflected glare of lights from the city. There was a fine mist in the street and, in it, the lighted dome of the Capitol looked to Booth like an apparition viewed through a frosted window. Still, the streets were crowded and a parade of government employees was making up and musicians were sounding their instruments. The gaiety, in fact, seemed enhanced by the hour, and the same revelers who, for days past, had vowed that they could not celebrate any more, were out again tonight drinking toasts to Lincoln, Grant, Stanton, Meade, McClellan, Sheridan, Ord, Sherman, Hooker, Farragut, Porter, and, when they got to thinking of Lee, they knew they were good and drunk.

Vice President Johnson was lying on his bed, clothed. He heard a knock on the door. He got up, adjusted his galluses, and opened it. The visitor was ex-Governor Leonard J. Farwell of Wisconsin. He said he was tired of sitting around the hotel. He needed someone to talk to and he wanted to stop in for a minute or two. Johnson turned up the table lamp and the two men, both lonely, talked peace and politics and Farwell said that he had a ticket to go to Ford's Theatre tonight. It had been announced, he said, that Lincoln and Grant would be there.

The Vice President did not rise to any faintly implied invitation to come along. He told the governor that he was tired— very tired—and he was going to get to bed early.

At Mrs. Surratt's house, John Holahan awakened from a nap and looked at his watch. He reminded Mrs. Holahan that, tonight, the employees of the arsenal were putting on a torchlight procession from the Capitol to the White House and that it would be a good idea to go down on the Avenue and watch it. Mrs. Holahan was not interested. She could think of things that had to be done right here in the house without getting dressed to stand out on a wet night watching parades.

Holahan washed his face in a basin, put his coat on, and left. He walked down Sixth Street to the Avenue, then over to Seventh, and stood in front of Seldner's Clothing Store and waited for the parade. The big Irishman was moved by the ruddy flames of the torchbearers and the music and the grinning faces and, when the parade had passed, he decided that he might as well go to Ford's Theatre and see Lincoln and Grant.

He got as far as Eighth Street and D when he remembered that this was Good Friday. Holahan was Catholic. He decided to go home. Had he gone to the theater, he stood an excellent chance of purchasing one of the few remaining seats in the theater, the first seat in Row D of the dress circle—a few short steps from the little white door. Had he been seated there, he would have seen anyone passing by en route to the President's Box and, had he recognized an idol, Holahan would have engaged him in conversation beyond doubt. What would have happened no one knows, but Holahan was a big man, and a strong one.

In the White House, the President worked. Speaker Colfax was in the office and he was explaining to Lincoln that he had scheduled a summer trip to the Rocky Mountains and California but he wanted to cancel it at once if the President decided to call a special session of Congress. Lincoln shook his head. He had not decided to call a special session, and he did not expect to call one. Let the Speaker go on with his trip. In fact, right now the President would write out a few remarks about the important part the people of the West Coast would play in the coming peace with their gold and silver mines.

Colfax, after a pause, said that many people had been worried about the President exposing himself to violence in Richmond. Lincoln said: "Why, if anyone else had been President and gone to Richmond, I would have been alarmed too; but I was not scared about myself a bit."

The two got to talking about the theater party tonight and Lincoln said that General and Mrs. Grant had intended to accompany them, but that Grant had been in a hurry to see his children and had taken the night train to Philadelphia. Would the Speaker like to join the party?

Mr. Colfax declined, with thanks.

Out in the front corridor, John F. Parker walked up to the big front doors and chatted with his friend Tom Pendel. Mr. Pendel had been one of the four original presidential guards, but had been advanced to the post of White House doorman. Parker had filled the vacancy.

"John," said Pendel, "are you prepared?"

Parker looked confused. Prepared for what? Alfonso Dunn, the second doorman, spoke up: "Oh Tommy, there is no danger."

"Dunn," said Pendel, "you don't know what might happen." He had always been a careful policeman and he knew that Parker wasn't. "Parker," he said, "now you start down to the theater and be ready when he reaches there." He held one of the doors open. Parker, half amused, half blank, walked out into the night. "And you see him safe inside," said Pendel.

An usher brought word to Mr. Lincoln that former Congressman George Ashmun was waiting outside. The Congressman did not have an appointment, but what he had to see the President about would take but a minute or two. The President studied his timepiece. It was 7:50.

"All right," he said. "Show him in." He turned to Colfax smiling, and asked the Speaker to please wait outside for a moment. The hour was late, but the President could hardly refuse Mr. Ashmun "a minute or two" because Ashmun had presided over the 1860 convention which had nominated Lincoln. In Congress, he had been fairly loyal to Lincoln policies.

Now Ashmun sat and said that a client of his from his native Massachusetts had a sizable cotton claim against the government and Ashmun wondered if Mr. Lincoln would appoint a commission to examine into the merits of the case. The President listened with unconcealed irritation.

"I have done with commissions," he said angrily. "I believe they are contrivances to cheat the government out of every pound of cotton they can lay their hands on."

The ex-Congressman gasped. "I hope," he said, "that the President means no personal imputation."

Lincoln began to feel bad. "You did not understand me, Ashmun. I did not mean what you inferred. I take it all back."

Ashmun was mollified.

8 p.m.

It was 8:05 when Mrs. Lincoln, in pretty bonnet with tiny pink flowers, and low-necked white dress, stood in the office doorway pulling on gloves and said:

"Would you have us be late?"

The President fumbled for his watch, and remembered that he had asked Colfax to wait outside. He asked Ashmun if he would mind coming back again in the morning, when he would have plenty of time. Lincoln had momentarily offended a friend and now he wanted to lean over backward to swallow the honest words he had uttered.

Ashmun said that he would have time in the morning. The President took a card from his vest pocket and wrote on it in a large, shaky hand:

Allow *Mr.* Ashmun & friend to come in at 9 A.M. to mor row
A. Lincoln
April 14, 1865

He got up, excused himself, got his silk hat, brushed his hair with his hand, and, with Congressman Ashmun, joined Colfax, Mrs. Lincoln and Noah Brooks on the front porch. In a last-minute afterthought, he told Colfax that Senator Sumner had a gavel which the Confederate Congress had used and which Sumner wanted to present to Mr. Stanton. "I insisted then that he must turn it over to you," Lincoln said. "You tell him for me to hand it over."

His mind was still far from the theater. He watched the foot-man help Mrs. Lincoln into the closed coach and he said to the assemblage: "Grant thinks we can reduce the cost of the army establishment at least half a million a day, which, with the reduction of expenditures of our navy, will soon bring down our national debt to something like decent proportions, and bring our national paper up to par, or nearly so, with gold—at least, so they think."

He was "unusually happy." That was the reaction of Noah Brooks and Tom Pendel and Colfax. Not because of the theater engagement, but rather because he was now President of the *United* States. So far as Ford's is concerned, Lincoln stated his feelings to Senator Stewart succinctly:

"I am engaged to go to the theater with Mrs. Lincoln. It is the kind of engagement I never break."

He waved to all and, as he stepped halfway into the carriage, Isaac N. Arnold came out of the velvet dark of the driveway, yelling to the President and waving his arms. Arnold was another old friend. He had neglected his own seat in Congress to preach Lincoln up and down the state of Illinois and, as a result, he had been defeated.

The President backed out of the carriage and listened to Arnold's confidential whisper, with head cocked. Then he said: "Excuse me, now. I am going to the theater. Come and see me in the morning."

Up on the box in front, Forbes, the presidential valet and footman, sat back and folded Mr. Lincoln's heavy plaid shawl over his arm. He nodded to Burns, the coachman. The horses stepped forward, and gravel crunched under the wheels as the sixteenth President of the United States looked back at the White House and waved to his friends.

The conspirators held a final meeting.* Final agreements were reached. Booth would go to Ford's Theatre alone. He would ar-

---

* Where has always been a question. Most authorities believe that it was held in Paine's room at Herndon House, but Paine gave his room up at 3 P.M. The meeting was probably held outdoors, on horseback.

rive at nine or later and would assess the situation until 10:15, when he would strike. Afterward, he would head directly for the Navy Yard Bridge and, if the other actions were properly synchronized with his, all of the parties should meet at the bridge. Failing that, each party should proceed to Surrattsville and wait at the tavern. Lloyd had the guns and binoculars and, after a brief rest, the party would proceed directly to Port Tobacco and, once on the Virginia shore, head south.

Herold would guide Paine to Seward's home and lend whatever assistance might be necessary to dispatch the Secretary of State. In the event that no help was necessary, Herold would wait on the street with the horses, and guide Paine to the Navy Yard Bridge afterward. Atzerodt was to knock on the door of the Vice President's room—also at 10:15—and, when he answered the knock, shoot him. If someone else answered the knock, Atzerodt would push that person aside or, meeting resistance, shoot that person, and then find the Vice President and kill him. In leaving the Seward home, Booth thought it would be wise if Davey led Paine northward on Madison Place, giving any pursuit the illusion that the conspirators were headed in that direction, and then make a right turn and again a right turn so as to come down Fifteenth or Fourteenth Street and then turn easterly toward the bridge. If each of the parties struck at almost the same moment, an alarm spreading from one place would be met by an alarm coming in the opposite direction, and this was bound to confuse the government into thinking that the city was full of assassins. Timing was important.

The weakest point in the final plan was the matter of where to run to after Surrattsville. There was no longer a Confederacy to flee to for succor, assuming that the leaders of the South would lend themselves to assassination in the first place. John Wilkes Booth knew this. He was aware of current events and he prided himself on his knowledge of politics.

He might not have known that, now that the war was lost to the Confederacy, the literate citizens of the Southern states looked to Lincoln for a merciful peace, and would be horrified by his death and the ascension to power of the stern Andrew Johnson.

Still, although there was no friendly Confederacy to flee to, neither Paine nor Atzerodt nor Herold questioned the final plan of deed and escape. Herold suggested that he and Paine could get into Seward's house with less trouble if they claimed to have a prescription from a drugstore. He could arrange a small package and he knew that Seward's physician was Dr. Verdi. They could say that they had a prescription from Dr. Verdi and that the dosage had to be explained to the Secretary of State in person.

Booth thought about it. He agreed that it was a good idea. He closed the meeting by announcing that he had written a letter to a newspaper, explaining the high patriotic motives of all concerned in the scheme, and that they were "all in this thing together." He had taken the liberty of adding their names to his. By indirection, this canceled all ideas of deceit on the part of the conspirators. It would be of no use now to think of *not* participating in the murders because Booth had committed them to the overall plot and, if one government official was killed, all would hang.

The letter locked the door behind them.

The leader asked Paine, with slight levity, how he liked the one-eyed horse he was riding. Paine said that the horse rode hard. Booth shook hands all around, and wheeled his little mare away. The band broke up into missions.

The President's carriage turned north at Fifteenth Street and east on H Street. It pulled up in front of the home of Senator Ira Harris. Forbes got down off the box and rang the bell. Miss Harris and Major Rathbone, flushed with excitement at the unexpected honor, came out and stepped into the carriage. They faced the Lincolns, riding backward. The major, who affected muttonchop whiskers and a walrus mustache, was tall and slender and dignified. He was not in uniform, and he was not armed. The ladies talked animatedly on the short ride down Fourteenth, over F, and into Tenth Street.

Burns walked the horses as they approached the theater. Around the wooden ramp in front of Ford's, a few soldiers on leave sat or stood, gawking, waiting to see the President and General Grant. Inside, Act One was being played to an almost capacity house, although a small queue of patrons still waited for tickets at the box office. In the row of small houses across the street, people stood at windows, with curtains back, waiting to see the Chief Executive. Negro coachmen in cocked hats sat atop glittering carriages parked up and down both sides of Tenth Street. A policeman stood in front of the main entrance to the theater, keeping the pedestrians moving. Two cavalrymen, who had been riding behind the President's coach, now turned and swung back up Tenth Street to return to their units. They would come back after the last curtain, to escort the President home.

Against the wall of the main entrance, John F. Parker leaned. He had walked over from the White House, had studied the playbills of the theater and had looked at the crowds going in. A gusty wet breeze had come up, and Parker had gone into the theater and upstairs to the State Box. He saw the chair that he was to occupy, outside the white door, and he knew that, from there, he could not see the play; his back would be to the stage.

He went into the blind corridor behind the boxes, and opened the door to Box 7 and studied the layout. Everything looked all right. He came out, closed both doors, and went back downstairs. As he left, he saw Professor Withers lift his baton, heard the first soft strains of the overture, and saw two

Negro boys in red satin breeches come out through the flap in the curtain, and begin to lift both sides of it as they walked to the far sides of the stage. Many of the patrons looked up to the State Box and saw that it was empty. The show was on.

It was 8:25 when Lincoln arrived. Burns pulled up to the wooden ramp to protect the ladies' dresses and the men's boots from the mud. Forbes jumped down and helped the party to alight. A small knot of people gathered, and the policeman ordered them to make way. Parker did not move from his position. His duty, as he saw it, was partly done. He had examined the premises and found them to be free of danger to the President. Now he would lead Mr. Lincoln, by a pace or two, into the theater. Later, he would sit outside the corridor and, in time, see that the President got back into his carriage. At twelve midnight, Parker would be on his way home.

As the party walked into the theater, Burns swung the team around, and pulled up and parked north of the theater. Forbes, with the shawl, followed Major Rathbone, last of the group to alight, into the theater.

Mr. Buckingham, the ticket taker, bowed deeply to one and all. Parker led the way up the stairs. Onstage, Laura Keene, as Florence Trenchard, was in a scene with Lord Dundreary. He had mentioned a window draft, a medical draught and a draft on a bank.

"Good gracious!" said Miss Trenchard. "You have almost a game of draughts."

Lord Dundreary laughed hysterically.

"What is the matter?" said Miss Trenchard.

"That wath a joke," said Lord Dundreary. "That wath."

As the audience chuckled, patrons in the dress circle began to stand and applaud and, onstage, Miss Keene stopped the action and applauded vigorously. The entire theater stood. The President of the United States was leading his party down the side aisle, through the white door, and into the State Box.

Miss Keene ad-libbed: "The draft has been suspended."

"I can't see the joke," said Dundreary.

"Anybody," said Miss Keene, "can see that."

Professor Withers raised his baton and the band swung into "Hail to the Chief." Miss Harris and Major Rathbone could be seen taking their positions toward Box 8. Mrs. Lincoln was next—moving toward the front of the theater. Dimly, the figure of the President could be seen, partly hidden by drapes.

The band finished with a flourish, and everyone sat to prolonged applause. Forbes, the valet, sat to the rear of the box on a straight-backed chair and leaned over to whisper to the President about the shawl. Mr. Lincoln said that he did not want it at the moment. Forbes remained a little while, and then he left to go out and sit with the coachman.

Parker sat outside the corridor and looked at the faces in the dress circle and saw two army officers come down the side aisle and take the last empty seats in Row D. Several times Parker arose from his seat and peered around the edge of the wall at the action onstage.

The President was becoming absorbed in the play as, on H Street, Mrs. Surratt stepped out of the rig. Louis Wiechman said that he would return the horse and buggy and the widow said that, when he came back, she'd put up something to eat.

Mr. Stanton decided that he would go over and pay a call on Seward before retiring. It was 8:30 and he could be back in a half hour. At The Old Clubhouse—the Seward residence—he found the secretary to be in pain. He could not bear to have the arm moved, even when his nurses wanted to change his position.

Mrs. Seward had come down from Auburn, New York, and the family had agreed to sit certain watches with Mr. Seward. The mother sat the early watch from 6 P.M. until 9 P.M. Two of the younger children, Miss Fanny, a sensitive young lady who dreaded to see anyone in pain, and Major Augustus Seward, took over the night watch. Miss Fanny sat with her father from 9 P.M. until 11 P.M. The young major was with his father the rest of the night. In addition, Secretary Stanton had assigned two convalescent soldiers to assist the family.

The Secretary of War tried to be cheerful, but the Secretary of State groaned so much that it is doubtful if he heard what was said. Stanton left.

In the War Department, telegrapher David Homer Bates worked alone. He sat under a small night light, reading. The telegraph key began to chatter and Bates picked up a pencil and began to write:

> *City Point, Va. April 14, 1865*
> *(Received 8:30 p.m.)*
> Hon. E. M. Stanton:
>   I send you the farewell address of Lee to his army, which I obtained a copy of at Appomattox Court-house just as I left there day before yesterday . . .
> E. B. Washburne.

Just before nine o'clock John F. Parker became bored. He got up, pushed his chair against the dress circle wall, and walked up the aisle and out of the theater. At the corner, he saw Francis Burns dozing in the driver's seat of the President's carriage.

"How would you like a little ale?" Parker said.

Burns awakened, and said it was a good idea. The two started down to Taltavul's when Forbes came out of the theater and joined them.

9 p.m.

The big lights outside of Ford's Theatre were haloed in mist. Dimly coachmen could be seen hunched deep in their coats, their horses sleek and patient. The same off-duty soldiers waited in front of the theater, hoping for another glimpse of the President at intermission.

Inside, there were 1,675 persons. At least one was in a romantic mood. This was the President. He noticed that Major Rathbone, watching the play, had taken Miss Harris's hand in his. So he reached and found Mrs. Lincoln's hand and held it at the side of the rocker. After a moment, when he did not let go, Mrs. Lincoln leaned close to her husband and whispered:

"What will Miss Harris think of my hanging on to you so?"

"Why," the President said, not taking his eyes from the stage, "she will think nothing about it."

He maintained his grip. When the lights went up after Act One, the people in the audience studied the colorfully decorated box. Three persons could be seen plainly; the fourth sat in shadow. Those who were in the State Box could study the audience, a most fashionable audience of handsomely dressed ladies and stalwart men, many in uniform.

In an orchestra seat, Julia Adelaide Shepard used the intermission, and the brightened light, to write a hurried note to her father back home.

Cousin Julia has just told me that the President is in yonder upper right hand private box so handsomely decked with silken flags festooned over a picture of

George Washington. The young and lovely daughter
of Senator Harris is the only one of the party we can
see, as the flags hide the rest. But we know "Father
Abraham" is there; like a father watching what
interests his children, for their pleasure rather than
his own. It has been announced in the papers that
he would be here. How sociable it seems, like one
family sitting around their parlor fire. . . . Every one
has been so jubilant for days . . . that they laugh and
shout at every clownish witticism, such is the excited
state of the public mind.

One of the actresses, whose part is that of a very
delicate young lady, talks of wishing to avoid the
draft, when her lover tells her "not to be alarmed, for
there is no more draft" at which the applause is loud.
The American Cousin has just been making love to a
young lady who says she will never marry but for love,
yet when her mother and herself find he has lost his
property they retreat in disgust at the left of the stage,
while the American cousin goes out at the right. We
are waiting for the next scene.

At Kirkwood House, Andrew Johnson went to bed. Governor
Farwell had gone off to the theater. Johnson's secretary, Browning, was out. There was no one to talk to, and he had read all
that he wanted to read. The night was damp; the bed felt warm.

The food was always plentiful and plain at Surratt House
and Louis Wiechman and Mrs. Surratt ate a late supper.
He talked with Anna Surratt, who, with young Honora Fitzpatrick, was in a teasing mood. The upstairs bell rang and
Anna left the basement and hurried upstairs. Five minutes
later, Wiechman* heard someone walking back down the

---

* At the trial of the conspirators, he said that Mrs. Surratt, not Anna, answered the
doorbell, and he implied that the mysterious caller was Booth.

front steps. Anna came back to the basement, but she volunteered no information about the caller and the big boarder was piqued.

Mrs. Surratt, with pride, brought out the letter from Canada and showed it to Wiechman, who read it and handed it back without comment. The ladies went upstairs to the sitting room for the evening—Mrs. Surratt was a fair pianist—but Wiechman begged off and said that he was tired. He was in bed at 9:45. The people of Surratt House were indoors for the night.

At the White House, S. P. Hanscom called. Mr. Hanscom was a small, persevering man, editor of the *National Republican,* a newspaper which was liberal, unreliable, gossipy, and tried to exude the aura of being the President's unofficial organ. Hanscom irritated most editors and all reporters by being so ingratiating in his dealings with Lincoln that, little by little, he was permitted to walk into the President's office without appointment, at any hour of day or night and, after light chatter with Lincoln, would go back to his office and write an entire column on the state of the Union and the conduct of the war.

Now Mr. Hanscom was surprised to find that, had he read his own newspaper today, he would have learned that the Lincolns were going to Ford's Theatre. He sat with a few members of the staff and gossiped. A sergeant came in from the War Department and said that he had a telegraph message for the President. It was sealed, and at once the staff assumed that it contained news of Johnston's surrender. It would have to be delivered to the theater at once.

Hanscom said that he was walking up toward Tenth Street and would be glad to deliver it. The message was given to him. The editor left at once and arrived in front of Ford's at intermission time. The front walk was crowded with patrons, out for a breath of air or a cigar. Hanscom looked for the President's carriage, found it near the F Street corner, but no one was in attendance, so he went back to the theater and on up the stairs to the dress circle. As he passed down the side aisle, he saw two uniformed officers of the United States Army on the edge of Row D, and he asked quietly where he could find the President. One pointed to the little white door.

Hanscom went there, and found the President's valet, Forbes, sitting against the wall. The editor said he had an important message for Mr. Lincoln and Forbes thumbed him inside. The message was delivered (when he entered Box 7, Miss Harris turned around in alarm) and Lincoln thanked Hanscom cordially. The editor left and went home.

The message was of small consequence. It read:

Richmond Va. April 14, 1865 11 a.m.
(Received 9:30 p.m.)
President of the United States:
    Mr. R. M. T. Hunter has just arrived under the invitation signed by General Weitzel. He and Judge J. A. Campbell wish a permit for their visit to you at Washington, I think, with important communications.

E. O. C. Ord,
Major-General.

Both men had been prominent in the Confederate States of America, Hunter as a Secretary of State and Campbell as a Justice. The matter of whether Lincoln would permit them to come to Washington City could wait until morning.

At The Old Clubhouse, Dr. Verdi, Seward's physician, paid a short visit to his patient. The few visitors in the room excused themselves and waited outside in the third-floor hall. When the doctor emerged, he said that the patient was doing as well as could be expected, and pain had to be expected; sometimes, unremitting pain. He had left instructions with the sergeant about a sedative, and the Secretary of State was now dozing and he would suggest no more visitors tonight.

At 9:30, a mare picked her way lightly through the dark alley behind Ford's Theatre. She was sure-footed and her dainty feet rang on the stones as she approached the single light at the stage door.

Booth dismounted and shouted "Spangler!" He held the reins forward and waited. The only answer he got was the surf roar of laughter from inside the red brick building. Up high in the flies, John Miles, a Negro, heard the actor and, through the tall, almost cathedral back window, could see John Wilkes Booth standing outside with his horse. Booth called twice more. Miles, looking downward through the flies, saw Ned Spangler, who shifted scenes on the same side of the theater as the President's Box, leave his post and run to the door.

Still looking, Miles could see Booth talking to Spangler and tender a bridle rein. In the pantomime, he could see Spangler gesturing toward the theater, probably pleading that he was too busy to take care of the horse. Miles saw Spangler come back into the theater, go into the Green Room, and come out with Johnny Peanut. Peanut went out back, sat on the stone step, and held the mare.

Booth came in, removing gauntlet gloves, bowing and smiling to fellow actors, and whispered to an actor in the wings. The actor shook his head and pointed to the tunnel. Booth could not cross backstage at that time. In the wings, the conspirator tried to look across the stage to the President's Box, but Miles could see by the way he shaded his eyes that the powdery haze prevented him from seeing much.

At this moment, or a moment close to this one, President Lincoln told Mrs. Lincoln that he felt a chill. She wanted to get the comfortable shawl that Forbes had brought, but the President stood and put on a black coat instead. He sat in his rocker, and, looking across the stage to the wings opposite, he could no more see the assassin through the hanging lights than the assassin could see him.

Booth listened to the lines of the actors, lines which he could mouth with them. The responsible utility man, J. L. Debonay, stood beside him, hands jammed in his pockets, watching the action, and Booth asked if he could cross behind the set.

"No, Mr. Booth," said Debonay. "The dairy scene is on. You will have to go under the stage."

Booth went down in the subterranean passage. Overhead, he could hear the creak of the boards, the mumble of actors, the shrill laughter of women in the audience. He came up on the other side of the stage, peeked out at the packed house, and went out through the side alley to Tenth Street.

He had time.

Down on E Street, Atzerodt decided to again pick up the horse he had rented in one stable and boarded in another. He walked into Tim Naylor's place, across from Grover's Theatre, and asked John Fletcher for his mare. Atzerodt had been to this stable several times with David Herold, and Fletcher, who had a chronic fear of having horses stolen from him, didn't like either one of these men. Still, this mare didn't belong to Fletcher, so he brought her out, saddled and bridled her, and said she "looks kind of scarish."

Atzerodt was grinning and perspiring. "Will you have a drink with me?" he said.

Fletcher took a hitch in the bridle around a stable post and said, "I don't mind if I do." They walked a few doors down to the Union Hotel and Atzerodt had a whiskey and Fletcher drank a big schooner of beer. On the way back, Atzerodt, feeling the camaraderie engendered in buying a man a drink, said: "If this thing happens tonight, you will hear of a present." Fletcher, who had no idea of what might happen tonight, said nothing. He had already made up his mind that this man was drunk.

At the stable, he helped the customer to mount. Fletcher took five dollars, an exorbitant amount for boarding a horse for a few hours, and threw in some advice free: "I would not like to ride that mare through the city tonight," he said. "She looks so skittish."

Atzerodt settled himself in the saddle. "Well," he said, "she is good on retreat."

"Your friend," said Fletcher, "is staying out very late with our horse."

Atzerodt slapped his heels into the horse and said: "He'll be back after a while." The mare moved out onto E Street in a slow, biased trot. Without guidance, she turned east and Atzerodt almost fell out of the saddle. He had wanted to turn west, but had forgotten to guide the animal. He forced her to make a big turn in the street, then headed her to the point where E slices into Pennsylvania Avenue, then he turned left.

Fletcher stood in the stable doorway puffing on a pipe. He had a shrewd thought. Atzerodt might lead him to Herold. The two men were friends. So, leaving a stable full of animals alone, Fletcher started off on foot after Atzerodt. The trail led to Twelfth Street on the Avenue. The stable foreman saw Atzerodt dismount, hitch his horse, and go into Kirkwood House. Fletcher waited. In a few minutes, the carriage maker came out, wiped his mouth on his sleeve, and remounted. He went off up Twelfth Street, in no apparent hurry, and Fletcher decided to go back to the stable and hunt for Herold later.

John Wilkes Booth stood in front of Ford's, studying the play-bills. One announced a benefit tomorrow for Miss Jennie Gourlay in *The Octoroon*. Another said that Mr. Edwin Adams would appear at Ford's for a limited engagement of twelve nights only.

Booth went into Taltavul's and asked Peter Taltavul to set a bottle of whiskey and some water before him. This was unusual, and Taltavul remembered it, because the actor usually asked for brandy. Booth drank. Farther down the bar, Burns the coachman and Forbes the valet had rejoined John F. Parker for a few more drinks. A man gaily intoxicated, lost in anonymity, lifted his glass to Booth and said: "You'll never be the actor your father was."

The conspirator smiled and nodded. "When I leave the stage," he said quietly, "I will be the most famous man in America."

10 p.m.

The night air cleared. The mists rolled away with theatrical speed and, in the gaps between the scudding clouds, the signals of far-off stars could be seen tapping blue dots and dashes. The moon was due to rise at 10:02 P.M. but the men at the Naval Observatory up beyond Rock Creek saw nothing in the east except the silvered edges of clouds over the Maryland shore.

The roisterers were still in the streets, and public singing was plentiful and cheap. At Lichau House, Mike O'Laughlin sang a flat baritone which most customers thought was good, and sad, or perhaps good and sad. In the freshly washed night air, the Capitol dome looked like a picture postcard and lights were on in many homes at an hour when most good families were in bed.

At Surratt House, the widow kissed Anna good night and began the job of turning off the kerosene lamps in the downstairs dining room and the upstairs sitting room, taking the last lighted lamp with her along the hall to the bedroom she shared with Honora Fitzpatrick. If she gave a thought to her son John, she thought of him in Canada, but, in reality, he was in northern New York State, in a small town where thousands of Southern prisoners were kept. He was on a final mission for the Confederate States of America.

George Atzerodt trotted his horse up Tenth Street again and he looked at Ford's Theatre as though fascinated. He saw the President's carriage and he saw off-duty soldiers lounging and he saw a few civilians on the sidewalk. He rode back to Kirkwood House to kill the Vice President but his feet carried him into the bar and he drank and looked at the clock and drank some more.

In the theater, the play was more than half over. The second scene of the third act had begun. President Lincoln, momentarily distracted from the action onstage, watched a portly officer come down the right-hand orchestra aisle. He knew the man. It was General Ambrose E. Burnside, an officer who did not believe that he was big enough to command the Army of the Potomac and, when Lincoln gave it to him, proved it. The President watched him come down front, split the tails of his uniform coat, and sit. Lincoln may have wondered what kept him so late. The presidential attention reverted to the stage.

Booth came out of Taltavul's and stood talking to Lewis Carland, the theater costumer. Mr. Carland was a sponge; he absorbed the moods of his friends. James J. Gifford, the stage carpenter, came out puffing a freshly lighted pipe and joined the conversation. A singer named Hess came down from F Street and asked what time it was. Someone looked at the lobby clock and said "ten." Hess returned a few minutes later and asked the same question. He was scheduled to go on, just before the last scene, and sing in concert with a young lady and another man "All Honor to Our Soldiers," the new song composed by Professor Withers.

Another man walked up from E Street to join the conversation. He was Captain William Williams of the Washington Cavalry Police. The captain was an admirer of John Wilkes Booth. He invited his idol into Taltavul's for a drink, but Booth looked at his watch, and declined with thanks.

"Keene," he said, "will be onstage in a minute and I promised to take a look for her."

He bowed and left the group and walked in the main entrance to the theater. Absentmindedly, John Buckingham, ticket taker, held out his hand, and Booth said, in mock shock: "You will not want a ticket from me?"

Buckingham laughed and bowed. "Courtesy of the house," he said. The actor looked at the lobby clock. It read 10:07. He saw Buckingham chewing, and borrowed a bite of tobacco. Buckingham said that, if Mr. Booth did not mind, he would like to introduce a few friends. The actor winked, and said: "Later, John." He turned and bounded up the stairs to the dress circle.

In Boston, his brother was, on this night, playing the part of Sir Edward Mortimer and, with a declaiming sweep of his hand, moaned: "Where is my honor now?"

Here in the dress circle, a man and a little girl were disappointed. James Ferguson, restaurateur, occupied the extreme left seat in the front row solely to see Abraham Lincoln and General Ulysses S. Grant. He did not want to see the play. He had brought the little neighbor's girl along because she too had an appreciation of historical figures which matched his. With her own eyes, she wanted to see the President of the United States and the man who had won the war.

All evening long, he had studied the State Box and the right-hand aisle. The President was in the box but, except for one brief moment when he had leaned forward to look down in the orchestra, they had not seen him. The general was not present and Mr. Ferguson kept telling the little girl that Grant was sure to be along at any moment. Now he saw a figure move down the right-hand aisle and he squeezed the little girl's hand. She followed his glance and saw a man step down the broad steps with easy grace. Ferguson shaded his eyes against the glare of the stage lights and, after a look, smiled sadly and said that it wasn't General Grant after all; it was a famous actor named Booth.

Almost as though to assuage the disappointment, James Ferguson noticed that, at the same time, President Lincoln was leaning forward in the box, with his left hand on the ledge, looking at the people below. It was the first time that Ferguson had seen Lincoln come into plain view, and he nudged the little girl and pointed. She looked steadily, and nodded. For the first time in her life she had seen, with her own eyes, the President of the United States.

John Wilkes Booth, slightly ahead of schedule, came down the dress circle steps slowly. He heard the lines onstage and he knew that he had about two minutes.

Asa Trenchard walked onstage and Mrs. Mountchessington said: "Ah, Mr. Trenchard. We were just talking of your archery powers."

Asa, who was played by Harry Hawk, was a slender drawling Yankee.

"Wal," he said, "I guess shooting with bows and arrows is just about like most things in life. All you have got to do is to keep the sun out of your eyes, look straight, pull strong, calculate the distance, and you're sure to hit the mark in most things as well as shooting."

Booth looked down at the little white door and saw the empty chair. Confused, he looked at patrons sitting in dress circle seats as though wondering which one was the President's guard. He saw the two army officers and he moved by them. For the first time, he realized that he was going to get into that box with no trouble; no challenge; no palaver; no argument; no fight; no stabbing. He was going to be able to walk in as though Lincoln had been expecting him.

He walked down to the white door, and stood with his back to it. He studied the faces nearby, men and women, and he saw some of them glance briefly at him. A real wave of laughter swept the theater and attention reverted to the stage.

Mrs. Mountchessington had just learned that Asa Trenchard was not a millionaire.

"No heir to the fortune, Mr. Trenchard?"

"Oh, no," he said.

"What!" young Augusta shrieked. "No fortune!"

"Nary a red," said Asa brightly. "It all comes from their barking up the wrong tree about the old man's property."

Now was the time. Booth knew that, in a few seconds, Asa would be alone on the stage. He turned the knob, pushed the door, and walked into the darkness. The door closed behind him. He found the pine board, held it against the inside of the door, and tapped the other end down the wall opposite until it settled in the niche he had carved for it. Pursuit could not come from that direction. Nor interference.

He moved toward the door of Box 7 in the darkness. A tiny beam of yellow light squeezed through the gimlet hole in the door and made a dot on the opposite wall.

Wilkes Booth could still hear the actors faintly. Mrs. Mountchessington had just said: "Augusta, to your room!"

And Augusta said: "Yes, Ma. The nasty beast!"

"I am aware, Mr. Trenchard," said Mrs. Mountchessington in her frostiest tone, "that you are not used to the manners of good society—"

The conspirator crouched and pressed his eye against the gimlet hole. What he saw was clear. The high back of the horsehair rocker was in plain view and the silhouette of a head above it. He waited. Three persons were on the stage. In a matter of seconds, Augusta would be offstage, followed by her irate mother. That would leave Harry Hawk (as Trenchard) alone and he would begin to drawl: "Don't know the manners of good society, eh? . . ."

Booth kept his eye to the gimlet hole. The head in front of him barely moved. The universe seemed to pause for breath. Then Trenchard said: "Don't know the manners of good society, eh?" Booth did not wait to hear the rest of the line. The derringer was now in his hand. He turned the knob. The door swung inward. Lincoln, facing diagonally away toward the left, was four feet from him. Booth moved along the wall closest to the dress circle. The President had dropped Mrs. Lincoln's hand and there was a little space between their chairs. The major and his Clara were listening to the humorous soliloquy of the actor onstage:

"Wal, I guess I know enough to turn you inside out, you sockdologizing old mantrap!"

The derringer was behind the President's head between the left ear and the spine. Booth squeezed the trigger and there was a sound as though someone had blown up and broken a heavy paper bag. It came in the midst of laughter, so that some people heard it, and some did not. The President did not move. His head inclined toward his chest and he stopped rocking.

Mrs. Lincoln turned at the noise, her round face creased with laughter. So did Major Rathbone and Miss Harris. A chrysanthemum of blue smoke hung in Box 7. Booth, with no maniacal gleam, no frenzy, looked at the people who looked at him and said, "Sic semper tyrannis!" It was said in such an ordinary tone that theatergoers only fourteen feet below did not hear the words.

The conspirator forced his way between the President and his wife. Mrs. Lincoln's laughter dissolved in confusion. She saw the young man towering above her, but she did not know who he was or what he wanted. The major saw the cloud of smoke and, without understanding, jumped up and tried to grapple with the intruder. Booth dropped the derringer and pulled out his knife. The major laid a hand on his arm and the assassin's arm went high in the air and slashed down. Rathbone lifted his left arm to counter the blow, and the knife sliced through his suit and flesh down to the bone.

The assassin moved to the ledge of the box and the major reached for him with his right arm. Booth shoved him and said loudly: "Revenge for the South!" Mrs. Lincoln began to rub her cheek nervously. She glanced at her husband, but he seemed to be dozing.

Harry Hawk faltered in his lines. He looked up at the State Box indecisively. In the wings, W. J. Ferguson, an actor, heard the explosion and looked up at the box in time to see a dark man come out of the smoke toward the ledge. In the dress circle, James Ferguson and his little friend saw Booth climb over the ledge of the box, at a point near where Boxes 7 and 8 met at the picture of George Washington, and watched him turn his back to the audience and, by holding on with his arms, let himself down over the side.

As he dropped, he pushed his body away from the box with his right hand. This turned him a little and the spur of his right foot caught in the Treasury regiment flag. As the banner ripped, and followed him to the stage in tatters, the actor, by reflex, held his left foot rigid to take the shock of the fall, plus two outstretched hands. He landed on the left leg, and it snapped just above the instep. He fell on his hands, got up, and started to run across the stage to the left. He passed Harry Hawk and headed for the wings.

The audience did not understand. They watched the running actor, and he fell again. He stood and, as he got offstage, he was limping on the outside of his left foot; in effect, walking on his ankle.

Hawk, stupefied, did not move. His arms were still raised in half gesture toward the wings through which the women had departed. Laura Keene, in the Green Room, noticed that the onstage action had stopped and she came out in time almost to bump into Booth. She brushed by him, wondering what had happened to Harry Hawk. An actor stood in Booth's way and he saw a knife flash by his face.

A piercing scream came from the State Box. This was Mrs. Lincoln. Clara Harris stood and looked out at the people below and said "Water!" Major Joseph B. Stewart, sitting in the front row of the orchestra with his wife and his sister, got up from his seat and climbed over the rim of the stage. He was a big man, looking bigger in a pale fawn suit, and he got to his feet, rushed by Harry Hawk, and yelled "Stop that man!"

The conspirator hobbled to the back door, opened it, and shut it behind him. Johnny Peanut was lying on the stone step with the mare's bridle in his hand. Booth's face was snowy and grim as he pulled his foot back and kicked the boy in the chest.

He took the bridle and limped toward the animal. She began to swing in a swift circle as he tried to get his good foot up in the stirrup. When he made it, Booth pulled himself across the saddle, threw his left leg over, and was just settling in the saddle when Major Stewart came out the back door yelling "Stop! Stop!" He reached for the rein as Booth spurred the horse and turned out of the alley.

The course he chose was not up to F Street, where the gate would have to be unlatched. He swung toward the side of the T, out through Ninth Street, then right toward Pennsylvania Avenue. His job was to put that first mile between him and his pursuers; he must be ahead of the news he had created. So he spurred the little mare hard, and she laid her ears back and ran. The conspirator was in little pain. He knew that his leg had been hurt, but the pain was not great now. He leaned his weight on the right stirrup and sat with the left thigh half up on the saddle. The mare turned into Pennsylvania Avenue and headed toward the Capitol. To the right of the House wing, a moon two days shy of being full was showing.

At Capitol South, he passed another horseman, trotting in the opposite direction. The speed of the mare attracted the lone rider's attention. As Booth turned into New Jersey Avenue, he slowed the mare. This was a shanty section, so dark that, unless the United States Government knew his escape route, no one would look for him here. At Virginia Avenue, he turned left, and was now close to the bridge.

When Booth swung away from the rear of Ford's Theatre, Johnny Peanut rolled in the alley, moaning: "He kicked me. He kicked me." Major Stewart turned to go back into the theater and was met by a rush of theater people coming out. Backstage, Jacob Ritterspaugh ran out of the wings and grabbed Ned Spangler by the shoulders.

"That was Booth!" he shouted. "I swear it was Booth!"

Spangler swung and smashed Ritterspaugh in the face.

"Be quiet!" he said. "What do you know about it?"

The audience began to buzz. Some of the men stood and began to ask others what did this mean. The people sensed now that this was not a part of the play and they felt vaguely alarmed. Major Rathbone pointed dramatically toward the dead wings and roared: "Stop that man!" Out of the State Box came a second scream, a shriek that chilled the audience and brought a large part of it to its feet. This again was Mrs. Lincoln. It had penetrated her mind that Mr. Lincoln could not be aroused. To the west, many farmers testified that, at this time, the moon emerged from behind clouds blood red.

In the orchestra, one man stood and brought to mouth the question everyone was asking: "For God's sake, what is it? What happened?" Miss Shepard, the letter writer, stood and saw that Miss Harris was leaning over the ledge of the box wringing her hands and pleading for water. Someone in the box, a man, yelled:

"He has shot the President!"

All over the theater, hoarse voices shouted, "No! No!" "It can't be true!" In a trice, Ford's resembled a hive immediately after the queen bee has died. The aisles were jammed with people moving willy-nilly. The stairs were crowded, some trying to get up to the dress circle, others trying to get down.

Some were up on the stage. Harry Hawk stood in stage center and wept. A group of men tried to force their way through the white door, but, the harder they pushed, the more firmly it held. James Ferguson, choking with horror, picked the little girl up and said that he would carry her out of the theater. Actors in makeup ran on the stage begging to know what had happened.

"Water!" Miss Harris begged from the box. "Water!"

Some of the patrons got out on the street and spread the word that Lincoln had been shot. The President, they said, is lying dead in the box inside. Tempers flared. A crowd collected. From E and F Streets, people came running. Many tried to get into the theater as others were trying to get out. Inside, a few women fainted and the cry for water could be heard from different parts of the theater.

Rathbone, soaked with blood, went back into the corridor and tried to open the door. He found the wooden bar and yelled for the men on the other side to stop leaning against the door. After several entreaties, he was able to lift the bar and it fell to the floor, stained with his blood. The major pleaded that only doctors be admitted. A short, handsome man in sideburns and mustache yelled from the rear of the mob that he was a doctor. Men pushed him forward until he got inside the corridor. He was Dr. Charles Leale, Assistant Surgeon of United States Volunteers, twenty-three years of age.

Someone, below the stage, turned the gas valve up and hundreds of faces were revealed to be in varying stages of fright and anger. On the street, a man shouted, "I'm glad it happened!" In a moment, he was scuffed underfoot, most of his clothes ripped from his body, and he was carried toward a lamppost. Three policemen drew revolvers to save his life.

In the State Box, President Lincoln's knees began to relax and his head began to come forward. Mrs. Lincoln saw it, moaned, and pressed her head against his chest. Rathbone asked Dr. Leale for immediate attention. "I'm bleeding to death!" he said. The blood had soaked his sleeve and made a pool on the floor. The doctor lifted Rathbone's chin, looked into his eyes, and walked on into the box.

Miss Harris was hysterical. She was begging everyone to please help the President. The doctor looked at her, then lifted Mrs. Lincoln's head off her husband's chest. The First Lady grabbed the hand of medicine and moaned piteously.

"Oh, Doctor! Is he dead? Can he recover? Will you take charge of him? Oh, my dear husband! My dear husband!"

"I will do what I can," the doctor said, and motioned to the men who crowded into the box behind him to remove her. She was taken to the broad sofa in Box 8, and Miss Harris sat beside her, patting Mrs. Lincoln's hand.

At first, Leale thought that the President was dead. He pushed the shoulders back in the rocker so that the trunk no longer had a tendency to fall forward. Then he stood in front of the President and studied him from head to foot. With the attitude of one who knows that he will be obeyed, he said to the gawking men: "Get a lamp. Lock that door back there and admit no one except doctors. Someone hold matches until the lamp gets here."

These things were done, as Dr. Leale knew that they would be. He was the first person to bring order around the dying President. The eyes of the patient were closed. There was no sound of breathing. There was no sign of a wound. Men held matches and looked open-mouthed as Leale placed the palm of his hand under the whiskered chin of the President, lifted it, and then permitted it to drop.

In the crowd peering in from the corridor, he saw a few soldiers. "Come here," he said to them. "Get him out of the chair and put him on the floor." Half afraid, they did as he told them to. The body was relaxed. They placed it on the floor and stepped away. Leale was going to look for the wound. He was sure that it was a stab wound because, as he was passing the theater on his way back to the army hospital, he heard a man yell something about the President and a man with a knife. Further, he had seen that Major Rathbone sustained a knife wound.

Dr. Leale crouched behind Lincoln's head and lifted it. His hands came away wet. He placed the head back on the floor and men in a circle held matches at waist level as the doctor unbuttoned the black coat, the vest, unfastened the gold watch chain, and, while trying to unbutton the collar, he became impatient and asked for a pocket knife. William F. Rent had a sharp one, and Doctor Leale took it and slit the shirt and collar down the front.

He tore the undershirt between his hands and the chest was laid bare. He saw no wound. The doctor bent low, and put his ear to the chest. Then he lifted the eyelid and saw evidence of a brain injury. He separated his fingers and ran them through the patient's hair. At the back, he found matted blood and his fingers loosened a clot and the patient responded with shallow breathing and a weak pulse.

Onstage, men lifted another doctor into the box. This one was Dr. Charles Taft. He was senior to Leale, but he placed himself at Leale's disposal at once as an assistant. Leale lifted the body into a slumped sitting position and asked Dr. Taft to hold him. In the saffron flicker of the matches, he found what he was looking for. His fingers probed the edges of the wound and he pulled the matted black hair away from it. It was not a knife wound. The President had been shot behind the left ear and, if the probe of a pinky meant anything, the lead ball moved diagonally forward and slightly upward through the brain toward the right eye. Dr. Leale felt around the eye to see if the ball had emerged. It had not. It was in the brain.

Gently, he lowered the great head to the floor. He knew that Lincoln had to die. Leale acquainted Dr. Taft with his findings, and his feeling. He straddled the hips and started artificial respiration. His business was to prolong life—not to try to read the future—and so he raised the long arms up high and lowered them to the floor—up and back—forward and down—up and back—forward and down. For a moment, he paused. Rudely, he pushed the mouth open and got two fingers inside and pushed the tongue down to free the larynx of secretions.

Dr. Albert F. A. King was admitted to the box. Leale asked each doctor to take an arm and manipulate it while he pressed upward on the belly to stimulate the heart action.

A few soldiers started to clear the box of people. From onstage, questions flew up to the box. Mostly, they were un-answered. "How is he?" "What happened?" "Was he stabbed?" "Who did it?" "Is he breathing?" "Did anyone see who did it?"

For the first time, someone uttered the name of John Wilkes Booth. The name moved from the stage down into the orchestra, was shouted across the dress circle and out of the half-empty theater into the lobby and cascaded into Tenth Street. "Booth!" "Booth did it!" "An actor named Booth!" "The management must have been in on the plot!" "Burn the damn theater!" "Burn it now!" "Yes, burn it!" "Burn!"

Grief spirals to insanity.

Dr. Leale sat astride the President's hips and leaned down and pressed until these strangers met, thorax to thorax. Leale turned his head and pressed his mouth against the President's lips, and breathed for him in a kiss of desperation. Then he listened to the heart again and, when he sat up, he noticed that the breathing was stronger. It sounded like a snore.

"His wound is mortal," he said to the other doctors. "It is impossible for him to recover."

One of the soldiers began to get sick. Two others removed their uniform caps. A lamp arrived. Dr. Leale saw a hand in front of him with brandy. He dripped a small amount between the bluish lips. Leale watched the Adam's apple. It bobbed. The liquid had been swallowed and was now retained.

He paused in his labors to wipe his face with a kerchief. "Can he be removed to somewhere nearby?" Leale said.

"Wouldn't it be possible to carry him to the White House?" Dr. King said.

"No," Dr. Leale said. "His wound is mortal. It is impossible for him to recover."

On the couch, Mrs. Lincoln sat quietly, rocking slightly. Miss Laura Keene had come into the box and was now sitting with her and with Miss Harris. All three heard Dr. Leale's words, but only Mrs. Lincoln seemed not to comprehend. She sat between them, rocking a little and looking across the theater at the other boxes.

Miss Keene came over, and asked the doctor if she could hold the President's head for a moment. He looked at her coldly, and nodded. She sat on the floor and placed his head on her lap.

"If it is attempted," said Leale, still thinking about the White House, "he will be dead before we reach there."

Dr. Taft asked an officer to run out and find a place nearby—a suitable place—for President Lincoln. He called four soldiers to carry the body—at first it was decided to try seating the body in the rocker and carrying it that way—but Leale said that there were too many narrow turns and besides, it would not hurt him to be carried as long as the open wound was downward.

Four men from Thompson's Battery C, Pennsylvania Light Artillery, drew the assignment. Two formed a sling under the upper trunk; the other two held the thin thighs. Dr. King held the left shoulder. Dr. Leale followed behind and held the head in cupped hands. Miss Keene sat, oblivious to the dark stain on her dress, watching. At the last moment, Leale decided that headfirst would be better and he walked backward with Lincoln's head in his hands, his own head twisted to see ahead.

"Guards!" he yelled. "Guards! Clear the passage!"

From somewhere, a group of troopers came to life and preceded the dismal party, shoving the curious to one side. "Clear out!" they yelled at one and all. "Clear out!"

At the head of the stairs, Leale shouted orders as the party began the slow descent. Ahead, they could hear the cries of the crowd in Tenth Street. Downstairs in the lobby, a big man looked at the great placid face, and he blessed himself. Tenth Street was massed with humanity as far as the eye could see.

A short paunchy captain of infantry impressed more soldiers to duty and ordered them double-ranked to precede the body. He drew his sword and said: "Surgeon, give me your commands and I will see that they are obeyed." Leale looked at the houses across the street, private homes and boarding-houses, and asked the captain to get them across.

For the first time, the crowd saw the shaggy head and the big swinging feet. A roar of rage went up. Someone in the crowd yelled "God almighty! Get him to the White House!" Leale shook his head no. "He would die on the way," Leale said. Men in the crowd began to weep openly. The little party pressed through, inch by inch, the faces of the mob forming a canopy of frightened eyes over the body. The crowd pressed in ahead, and closed in behind.

The paunchy captain swung his sword and roared: "Out of the way, you sons of bitches!"

The night, now, was clear. The mist gone. The wind cool and gusty. The moon threw the shadow of Ford's Theatre across the street.

Every few steps, Leale stopped the party and pulled a clot loose. The procession seemed to be interminable. When they got across the street, the steady roar of the crowd made it impossible to hear or to be heard. Leale wanted to go into the nearest house, but a soldier on the stoop made motions that no one was home and made a helpless pantomime with a key. At the next house toward F Street, Leale saw a man with a lighted candle standing in the doorway, motioning. This was the William Petersen house at 453 Tenth Street. Mr. Petersen was a tailor.

Lincoln was carried up the steps and into the house. Part of the crowd followed. The man with the candle motioned for the doctors to follow him. They moved down a narrow hall. To the right was a stairway going up to the second floor. To the left was a parlor, with coal grate and black horsehair furniture. Behind it, also on the left, was a sitting room. Under the stairway was a small bedroom.

Here, the President was placed on a bed. A soldier on leave, who had rented the room, picked up his gear and left. He was Private William T. Clark of the 13th Massachusetts Infantry. The room measured fifteen feet by nine feet. The wallpaper was oatmeal in character. A thin reddish rug covered part of the floor. There were a plain maple bureau near the foot of the bed, three straight-backed chairs, a washstand with white crock bowl, a wood stove. On the wall were framed prints of "The Village Blacksmith" and Rosa Bonheur's "The Horse Fair." The bed was set against the wall under the stairway.

It was too small for the President. Leale ordered it pulled away from the wall. He also asked that the foot-board be taken off, but it was found that, if that was done, the bed would collapse. The body was placed diagonally on the bed, the head close to the wall, the legs hanging off the other end. Extra pillows were found and Lincoln's head was propped so that his chin was on his chest. Leale then ordered an officer to open a bedroom window—there were two, facing a little courtyard—and to clear everybody out and to post a guard on the front stoop.

At the back end of the room, Leale held his first formal conference with the other doctors. As they talked in whispers, the man who had held the candle went through the house lighting all the gas fixtures. The house was narrow and deep, and behind this bedroom was another and behind that a family sitting room which spread across the width of the house.

Leale, in the presence of the other doctors, began a thorough examination. As he began to remove the President's clothing, he looked up and saw Mrs. Lincoln standing in the doorway with Miss Keene and Miss Harris. He looked irritated and asked them to please wait in the front room. The patient was undressed and the doctors searched all of the areas of the body, but they found no other wound.

The feet were cold to the touch up to the ankles. The body was placed between sheets and a comforter was placed over the top. A soldier in the doorway was requisitioned as an orderly and the doctors sent him for hot water and for heated blankets. They sent another soldier for large mustard plasters. These were applied to the front of the body, covering the entire area from shoulders to ankles.

Occasionally, the President sighed. His pulse was fortyfour and light; breathing was stertorous; the pupil of the left eye was contracted; the right was dilated—both were proved insensitive to light. Leale called a couple of more soldiers from the hallway, and sent them to summon Robert Lincoln, Surgeon General Barnes, Dr. Robert K. Stone, President Lincoln's physician, and Lincoln's pastor, Dr. Phineas D. Gurley.

The death watch began.

At ten minutes past ten, Lewis Paine and David Herold rode into Madison Place, across the street from the White House. They stopped in front of The Old Clubhouse. Three doors away, a sentry lounged in front of General Augur's personal quarters. Two gas lamps lost a battle with darkness. Paine dismounted, handed the reins to Herold. He repeated the name of the doctor "Verdi, Verdi" as though it was difficult to remember. He ordered Herold to wait for him and not to move from in front of the door.

He removed a bottle from his jacket pocket. David Herold, sitting his horse and holding the awkward blind one, watched Lewis Paine walk up to the front door and rap hard with the knocker. Through the chased glass panels light could be seen.

No one answered. Paine rapped again and waited. A shadow grew on the glass and the door opened: A young Negro in a white coat stood inside. This was William Bell.

"I have medicine from Doctor Verdi."

William reached for it. Paine pulled his hand away.

"It has to be delivered personally."

"Sir," said the boy, "I can't let you go upstairs. I have strict orders—"

The rare temper began to crumble. "You're talking to a white man," Paine said. "This medicine is for your master and, by God, I'm going to give it to him."

"But, sir . . ."

"Out of my way, nigger. I'm going up." Paine pushed his way into the big reception hall, and started up the stairs, William a step or two behind, pleading softly. Paine walked heavily. William Bell asked him to please walk easily.

"I'm sorry that I talked rough to you," Bell said.

"Oh," said Paine, at the top of the first flight, "that's all right."

On the top floor, Frederick Seward, Assistant Secretary of State, heard the commotion and the tramp of heavy boots. He had been in bed with his wife, and now he had put on a dressing robe and hurried out.

He saw Lewis Paine coming up toward him, and saw Bell directly behind him. Seward, angry, whispered a demand to know what the commotion was all about. Paine, stopping two steps below the top, whispered back that he had a prescription from Dr. Verdi and that this fresh nigger tried to stop him.

Seward held out his hand. He would see that the prescription was delivered. Paine shook his head. The doctor had told him twice to make certain that this medicine got into no other hands than those of the Secretary of State. If he was permitted to hand the bottle to Secretary Seward, he would leave at once. The young official didn't know whether to throw the messenger and his medicine out, or to reason with him. In his mind, Seward figured that this man was one of those dull mentalities who know no better than to obey orders literally.

"My father may be sleeping," he said. "I will see."

He went up to the front of the hall to a door on the left side. Until then, Paine had no idea where the Secretary of State might be. Now he knew. In a moment, Seward was back.

"You can't go in," he said. "He's sleeping. Give it to me."

"I was ordered to give it to the secretary."

"You cannot see Mr. Seward. I will take the responsibility of refusing to let you see him. Go back and tell the doctor that I refused to let you see him if you think you cannot trust me with the medicine. I am Mr. Seward and I am in charge here." The voice began to rise in tone. "He will not blame you if you tell him I refused to let you see him."

Paine hesitated. Then he said: "Very well, sir. I will go."

He turned and faced down the stairs. He pulled his pistol, whirled, and fired at the middle of Frederick Seward. The hammer clicked. There was no explosion. Paine jumped to the top step and, before Seward could lift his hand, the rare temper brought the butt of the gun smashing down on Seward's head. He fell and Paine bent over him, smashing again and again at head and neck.

Bell, halfway up the stairs, turned and ran down, screaming "Murder! Murder!" He ran down the second flight of stairs, still screaming the litanous word and out into the street. "Murder! Murder! Murder!" David Herold watched him. Quickly, the assassin's escort dismounted, tied Paine's horse to a tree, remounted, and galloped off. As he turned into Pennsylvania Avenue at Fifteenth, Booth was at the other end of the Avenue, turning into Capitol South.

Upstairs, Paine found that he had broken his pistol. He threw it at the unconscious man and drew a knife. He hurried to the front bedroom. When he pushed against the door, he found that someone was leaning against it. Paine moved back a step and crashed his weight against the panel. The door flew open and Paine fell, inside. The room was in darkness except for a slice of light from the hall.

The assassin got up, saw a moving figure, and slashed at it. He heard a man scream in pain. His duty was to kill the Secretary of State and he had no time for others, so he jumped on the bed and, when he felt the helpless figure beneath him, he struck with his knife again and again. He heard small moans and he lifted the knife once more, as high as he could. Someone jerked his arm from behind and he turned and found that, in the darkness, he was battling two men.

They were trying to pull him off the bed. No words were spoken. The Secretary of State, still conscious, had the presence of mind, when his assailant was removed, to roll off the bed onto the floor against the wall, even though he knew that he was falling on the broken arm. Paine hacked at the restraining arms around him. The three fell into tables and chairs and, when he felt himself free, Lewis Paine got up and ran out into the hall, yelling, "I'm mad! I'm mad!"

There he saw a young lady, in nightdress, screaming. At the same time, he saw another man coming toward him. This man was well dressed and seemed confused by all the noise. He walked toward Paine blindly. The assassin permitted him to come close, then raised his knife and plunged it into the stranger's chest up to the hilt. Mr. Hansell, State Department messenger, fell without uttering a word.

Paine hurried downstairs and out into the street. He looked

for Herold, and found that he had been deserted. He untied his horse, mounted, and, mopping his face, turned north toward H Street. He walked the horse and William Bell, seeing him, followed behind, cupping his hands and yelling "Murder!" Soldiers came running from Augur's sentry box. They passed the assassin, passed the Negro boy, who was pointing at Paine, and ran up the steps of The Old Clubhouse.

Bell was stubborn. He kept behind Paine until the assassin turned, annoyed, and spurred his shaggy-shanked horse into a trot. The boy still followed, for a block and a half. Then he stopped and hurried back to Mr. Seward's house.

The Seward home looked unreal. Hansell, barely conscious, was bleeding profusely and gagging on his blood. At the top landing, Frederick Seward lay curled on his side, in a coma. On the rug beside him was a broken pistol and a black felt hat—Paine's. A male nurse, Sergeant Robinson, was badly hurt and bleeding. Augustus Seward was injured, but not bleeding. Miss Fanny Seward, who had been smashed and knocked down when Paine had first entered the sick room, was unconscious on the floor. She was one of the "men" he thought he had been battling.

When William Bell got back to the house, Major Augustus Seward was standing in the doorway with a huge pistol in his hand. People came running from all over Lafayette Square. Little Bell tried to tell his story, and point to which way the man had gone, but no one had time to listen to him.

Paine outdistanced the shouts of murder and soon he found that he was in a maze of streets, all of them dark and lonely. He remembered that "Cap" had said to turn right, so he turned right. He trotted his horse and he walked his horse. After a half hour, houses became infrequent and he saw dark fields. In the moonlight, he saw some soldiers coming toward him, so he got off the horse and hid in a field. He was in the East Capitol section, but he didn't know it. He was also about a half mile from the Navy Yard Bridge, but he didn't know that either.

The streets of downtown Washington were alive with running people who shouted to darkened houses that assassins were at large and that the Secretary of State had been murdered in his bed. This wave of hysteria, as John Wilkes Booth figured, met an opposite wave which roared that the President had been killed in cold blood in Ford's Theatre.

The news reached different people in different ways. Major Eckert was standing before a mirror in his room, shaving, when a friend burst in and said that Seward had just been killed. Mr. Stanton was undressing for bed, having been serenaded by the arsenal band, when a soldier banged on the broken pull bell and then rapped on the door. Stanton heard the news about Lincoln and Seward, went back upstairs, and told Mrs. Stanton that it was humbug. He was getting into bed when more people came with the same wild news. He dressed and someone got him a hack and he hurried to Seward's home. Robert Lincoln had just arrived home, and was sitting with members of his father's staff, when the tragic news came. Surgeon General Barnes was homeward bound in his carriage and was passing Willard's Hotel when a cavalryman rode up, looked in, and advised the doctor to go to Ford's Theatre at once—the President had been shot.

Barnes ordered the driver to take him to his office at top speed. He wanted to get his instruments. He was packing them in a bag when a wild-eyed soldier burst in and said that Secretary of State Seward had been stabbed and to please hurry. The Surgeon General said that he had heard about the alarm, but that the man must have been confused because he had said it was the President, and the place was Ford's Theatre. Barnes went off to Seward's home. There he was dressing the wounds of Frederick Seward when a Negro hack driver pleaded his way up the stairs and begged Barnes to come at once to Tenth Street, the President of the United States was dying.

Robert Lincoln and John Hay raced to Tenth Street in a carriage. The President's oldest son did not believe the news. When the driver tried to turn off G Street into Tenth—a block and a half from the theater—a mass of humanity blocked the road and Robert Lincoln put his head in his hands and moaned. When soldiers tried to turn the carriage away, Lincoln, in anguish, said:

"It's my father! My father! I'm Robert Lincoln!"

With help, he got through on foot. When he saw his mother, in the parlor of Petersen House, he burst into tears.

In the Seward home, Nurse Robinson and Miss Fanny Seward had turned the gas up and Robinson had found the secretary on the floor between bed and wall. His eyes were open, staring into pools of his own blood.

Miss Fanny said: "Is my father dead?"

Robinson felt for a pulse and found none.

"He has no pulse," he said.

Miss Fanny threw up the front window and screamed "Murder! Murder!" Robinson tore the nightshirt open and listened for a heartbeat. He heard one, and it sounded strong. The Secretary of State whispered: "I am not dead. Send for a surgeon. Send for the police. Close the house."

The nurse lifted Mr. Seward and said: "Do not talk. It makes your bleeding worse." The patient was put back on the bed. Robinson got the twisted bedclothes off the floor and wrapped them around the secretary. Then he looked at the face on the pillow. With a cloth, he wiped the red mask off and saw two pulsing wounds, one on each cheek. The right cheek was slashed from ear to lip and hung in a flap over the lower jaw. From the side, Robinson could see the inside of Mr. Seward's mouth. The leather-covered iron brace around neck and jaw had saved the man's life.

Vice President Andrew Johnson heard a pounding on a door. He was half asleep, half awake. He heard it and yet he didn't hear it. It continued for some time. He got up, fumbling for the lamp beside his bed. Outside, former Governor Leonard J. Farwell said: "Governor Johnson, if you are in this room, I must see you." Johnson got the door open as Farwell was trying to peer over the transom.

The Vice President invited him in.

"Someone," Farwell whispered, "has shot and murdered the President."

Johnson, lighting the living-room lamp, swung around. He did not believe the news. Then he saw Farwell's wild, agonized expression and he ran to the man and they threw their arms around each other as though, without support, each would collapse. Farwell opened the door and peered both ways down the corridor. He rang for servants and asked for guards. One man was put inside the door and told to admit no one.

Someone knocked and Farwell, frightened to frenzy, refused to permit the door to be opened until he recognized the voice of a Congressman. The Congressman said that there were five hundred people in the lobby. Johnson emerged from his bedroom shoving his shirttails into his trousers.

"Governor," he said, "go back to the theater and find out how the President is."

In a little while, Farwell was back with Major James Rowan O'Beirne, Provost Marshal of the District of Columbia. There were a lot of people in Johnson's two rooms and the men were in such a state of excitement that they were ready to believe any idiocy. Farwell assured everyone that the President was dying; that Seward was dead, and that it was part of a gigantic plot to kill Johnson and all Cabinet ministers.

O'Beirne said that it was his opinion that Johnson should remain in his rooms with his friends. The Vice President bridled and insisted that his place was at the side of the President and that's where he was going. The Provost Marshal was opposed, but said that if Johnson had to do it, to wait until O'Beirne returned for him, when the excitement in the streets had died a little.

The stone had been dropped into the still pool. Now the wave began to ripple outward, evenly for the most part, and it spread to all parts of the city. Ella Turner, the prostitute who had loved Booth, heard of the deed and the name of the assassin. She went to her room, placed a photo of John Wilkes Booth under her pillow, and pressed her head into a rag soaked with chloroform.*

The news jumped from house to house, from street to street. In nightclothes, citizens gathered on the sidewalks, talking, and other citizens threw up the windows and demanded to know what the noise was for. In time, the wave reached reporter L. A. Gobright of the Associated Press. He was closing his office for the night—had the key in the door—when he heard the first wild rumor and, without waiting to check it, wired his New York office:

THE PRESIDENT WAS SHOT IN A THEATER TONIGHT AND PERHAPS MORTALLY WOUNDED.

Fifteen minutes later, all commercial telegraph lines out of Washington were dead and no further news got out of the city until 1 A.M.

In the home of Senator Conness, colleague Charles Sumner was chatting when a young man burst in and said, all in one breath: "Mr. Lincoln is assassinated Mr. Seward was murdered in his bed there's murder in the streets."

Sumner's reaction was: "Young man, be moderate in your

---

* She was discovered and revived.

statements. What has happened? Tell us." And when he heard it again, he did not believe it and he put on his cape and walked the short distance to the White House and said to the sentry: "Has Mr. Lincoln returned?"

"No, sir. We have heard nothing from him." Sumner went home.

The Chief Justice of the Supreme Court, Salmon P. Chase, heard the news at home and believed that it was mistaken. When he heard the same news from another source, Mr. Chase decided that he could not be of service to Lincoln, and stayed home and went to bed.

The Navy Secretary, Mr. Welles, was sleeping when Mrs. Welles awakened him with the news that Mr. Seward was dead. He dressed, hurried over to The Old Clubhouse. Stanton arrived at almost the same moment. They saw Frederick unconscious with two fractures of the skull; they saw blood and hysteria and anguish.

On the way downstairs, Welles admitted to the Secretary of War that he had not believed the news about Seward, but now he had seen it with his own eyes. He had heard that the President had been shot, but he did not believe that either.

"It is true," said Stanton. "I had a talk with a man who had just left Ford's Theatre."

"Then I will go at once to the White House."

"The President is still at the theater," said Stanton.

They had reached the downstairs reception hall and, as they talked, they noted the presence of many of Washington City's distinguished citizens.

"Then let us go immediately there."

"That is my intention. If you haven't a carriage, come with me."

Stanton ordered Quartermaster General Montgomery C. Meigs—the man who had built the Capitol dome—to take charge of the house and to clear it. General Meigs begged

Stanton not to go to the theater. He said that there was murder in the streets. Others took up the plea. Mr. Stanton, on one of the rare occasions of his life, hesitated.

Welles continued ahead and said: "I am going at once and I think it is your duty to go."

"Yes," said Stanton vaguely, "I shall go." But men hung on to his arms and Welles said he was wasting time and he went out and got into a carriage. Stanton followed and had one foot in the carriage when Major Thomas Eckert, on horseback, leaned down and begged him not to go. When Meigs saw that Stanton would go, he jumped into the carriage too, and yelled for a cavalry escort. The party started and Stanton jumped up in alarm and said: "This is not my carriage." Mr. Welles said that it was no time to argue about the ownership of a carriage. Stanton leaned out the window and invited Chief Justice David K. Cartter of the District of Columbia Court of Appeals to join them. The judge climbed up on the box beside the driver.

"The streets," said Welles, "were full of people. Not only the sidewalks but the carriage ways were to some extent occupied, all, or nearly all, hurrying toward Tenth Street."

The news had reached George Atzerodt and, when he heard it, the carriage maker was lost in panic. He galloped up F Street and, when a pedestrian yelled to him, Atzerodt hurried back to the stable where he had rented his horse and gave it up. On foot, he could not hope to join Booth in Surrattsville. He did not dare return to Kirkwood House. So he took a horse car for the Navy Yard. There he had a friend who owned a store. He would sleep on the floor of the shop.

A block away from Ford's, in a little hideaway saloon, John Matthews was drinking. The actor "who did not deserve to live" had finished his work at Ford's Theatre. He drank quietly, without conversation, until the news of the night slapped against the doors. Then his cloistered little world crumbled.

"What did he say?" John Matthews said to the bartender.

"He said the President was killed."

"I just left Ford's. The President was sitting in a box seat."

Matthews stood for a moment, mulling the word "killed," because that word would lead one to guess that there had to be a killer. John Matthews paid for his drinks and hurried out. The corner was full of people. A troop of cavalry, at dead gallop, ran through the crowd headed toward the theater. Matthews said to a man: "Who—who did it?"

A few people turned to look at him. The man shrugged. Another man said he had heard that an actor did it, somebody named Booth. John Matthews felt ill. He held his right hand against his chest, where the letter to the editor of the *National Intelligencer* reposed.

Matthews said that he did not know how he got back to his hotel, but he got there and he started a fire in the grate and then he sat and ripped open the letter and read it. What it had to say about the plot on Mr. Lincoln's life made him sick to his stomach. It was signed with the names of the men who were part of the plot, and Matthews felt that, by having possession of this incriminating letter, he too was a conspirator. He burned the letter in the grate, holding on to the flaming end of it until he had to let go.

A few of the words burned in his mind forever: "The moment has at last arrived when my plans must be changed. The world may censure me for what I am about to do, but I am sure that posterity will justify me. . . . John Wilkes Booth—Paine—Atzerodt—Herold."

On E Street, the celebration at Grover's Theatre was almost concluded when a man came into the theater and, standing behind the audience, shouted: "President Lincoln has been shot in his private box at Ford's. Turn out!" The audience buzzed. The actors paused. Corporal James Tanner, a bright young man who had mastered shorthand, stood and yelled: "Sit down! It's a ruse of the pickpockets!" His reasoning was that pickpockets were sitting in the audience and had arranged with one of their number to shout alarming news. Thus, in the rush to the exits, the pickpockets would fleece the crowd. The people listened to Corporal Tanner. They sat. Onstage, a young boy who had, a moment before, finished reciting a patriotic poem, came back and, his voice choking, announced that the news about President Lincoln was true.

There was no panic. Tanner and six hundred others were dazed. They went out onto E Street trying to convince themselves that Lincoln might be dead. There they learned that Seward had been murdered in his bed. Tanner and another soldier hurried to Willard's Hotel to get more details. They found a muttering mob of men who, in helpless rage, were ready to visit vengeance on anyone who said a disagreeable word. Tanner decided to go home. He had a room on the second floor of a house across the street from Ford's Theatre— next door to a family named Petersen.

Earlier, John Fletcher, still worried about Herold and the horse, stood in front of Willard's Hotel watching horsemen come and go. He saw a roan coming down in the darkness out of Fifteenth Street onto the Avenue. The horse was going fast but, as the rider approached the well-lighted hotel, he slowed to a trot. Fletcher assured himself that it was David Herold and he ran out into the street shouting:

"You get off that horse now! You've had that horse long enough!"

Herold, who was coming from Seward's home and was trying not to excite suspicion, pulled the horse away from Fletcher's outstretched hand and swung up Fourteenth Street toward F. When Fletcher saw the boy gallop off, he was certain that thievery was the object, so he ran back to the stable, saddled a dark horse, and hurried down E to Thirteenth, up Thirteenth to F, found that he had not headed the boy off, and could not see him ahead on F Street, and so he turned right again and went back to Pennsylvania Avenue. If Herold was going to steal the horse, he would head down to the Navy Yard Bridge and take the horse into southern Maryland. This constituted the first real pursuit of the conspirators. Mr. Fletcher was, at this time, about a mile and a half behind Booth and a half mile behind Herold.

At Third Street, the stable foreman turned around the south side of the Capitol and here he met a horseman coming the other way.

"Have you seen any horsemen going this way?" said Fletcher.

"Two," the stranger said. "Both very fast."

Fletcher spurred his horse. He was convinced that Atzerodt was the first man; Herold the second. They would take the stage road, down New Jersey Avenue to Virginia Avenue, then diagonally left to Eleventh Street, and then onto the bridge. It was dark and Fletcher was slow and careful. In retrieving a horse, he did not want to break the leg of another.

At the Navy Yard Bridge, Sergeant Silas T. Cobb was near the end of his tour of duty. Another hour and fifteen minutes, and he'd be headed back to the barracks, a few yards away. It was soft, safe duty, but it was also deadly dull. Cobb and two sentries patrolled the Washington City end of the long wooden bridge. They challenged all suspicious parties entering or leaving the city. At 9 P.M. every night, the bridge was closed. No one could leave Washington, and no one could come in. Of course, now that the war was over, no sergeant wanted to be severe on citizens, but a soldier had to be careful.

It was about 10:45 P.M. when Cobb heard hoofbeats in the darkness, approaching. A dark man with a black mustache came into the cone of light around the sentry box and one of the sentries put a hand on the rein and held the horse.

"Who are you, sir?" said Sergeant Cobb.

"My name is Booth."

"Where are you from?"

"The city."

"Where are you going?"

"I am going home."

Cobb looked the man over and walked around the horse. The man in the saddle had a fine smile.

"And where would that be?"

"Charles."

"What town?"

"No town."

"Come now."

"Close to Beantown, but I do not live in the town."

"Why are you out so late? You know the rules. No one is allowed past this point after nine o'clock. "

"That is new to me. You see, I had to go somewhere first, and I thought that I would have the moon to go home by."

Sergeant Silas T. Cobb studied the rider once more, and rubbed his chin.

"Go ahead," he said.

He stood in the light and watched the little mare pick her dainty way over the planks until he could no longer see her, but, far off in the middle of the span, he could hear the boom of the planking.

A few minutes later, Cobb heard a second horse. The sergeant came out of the sentry box and saw a young boy. The horse looked abused.

"Who are you?" said Cobb.

"My name is Smith."

"Ah, yes. Where are you bound for?"

"Home."

"What town?"

"White Plains."

"How is it that you are out so late?"

Herold gave a ribald reply.

Cobb brought him up close to the sentry box, took a good look, and told him to be off. Two conspirators were now reasonably safe in Maryland.

A few minutes later, a third horseman came into view. The sentry grinned at the sergeant.

"We're doing a good business tonight."

Before Cobb had a chance to ask questions, the third rider asked one.

"Tell me," said John Fletcher. "Did a man on a roan horse cross a few minutes ago? He had an English saddle and metal stirrups."

Cobb nodded. "Yes," he said. "He has gone across."

"Did he tell you his name?"

"Yes, Smith."

"Smith? Can I cross?"

"You can cross, but you cannot return back."

Fletcher thought it over. "If that is so," he said, "I will not go." He turned his horse and headed back into Washington City. The foreman was angry. He stopped at Murphy's Stable on the slight chance that his horse had been stabled there. It wasn't, but the stableman said that Fletcher ought to go back to his stable and remain there because the President had been killed and Seward was dying.

John Fletcher felt little interest in the news. He had lost a horse and he knew from past experience that old man Naylor would blame him. Fletcher resolved to get back to his stable, unsaddle his horse, and then walk up to police headquarters and register a complaint.

11 p.m.

Forty-five minutes had passed. From 10:15 P.M.—when it happened—until 11 P.M. nothing had been done. Official Washington was in a state of inert panic. The responsible men of government were much more concerned with preventing additional assassinations than they were with hunting and apprehending the perpetrators of the old ones. General Christopher C. Augur, in command of all the troops in the District, had sent mounted patrols out, but they were running helter-skelter. No one had sealed off the bridges and roads leading out of Washington. No one, in spite of the fact that John Wilkes Booth had been identified by dozens of persons in the theater as the murderer, sent a policeman to his room at the National Hotel.

Augur was willing to move, but he was afraid to do anything without instructions from Secretary Stanton. And the Secretary of War, for the moment, was frozen with fright. In his mind, he had the fixed notion that the South was making a last desperate bid for victory by instituting terror in the capital of the North. He saw this thing, not as the product of a pathetic band of four men, but as a broad Confederate plot which had only begun to unfold. In the light of his experience, Stanton was eminently justified. His department had contended, throughout the war, with real plots—high-level Confederate plots, if you please—and these included the Sons of Liberty in the Midwest, the attempt to burn New York City, the raid on St. Albans, Vermont. The mind of the Secretary of War had been conditioned to accept the fantastic in plots.

The man who moved first was Major A. C. Richards, Superintendent of the metropolitan Washington police. He had been in the audience at Ford's Theatre, an austere cop of military bearing, and he had seen and identified Booth as the man who had jumped from the President's Box. The major had left the theater, tried to locate the guard John F. Parker (who could not have been assigned to the White House without Richards's assent) and, not finding him, hurried back to police headquarters. He assembled the night detective squad, told them what had happened, and ordered them out at once to locate and bring in witnesses to the assassination. He reminded them that the Federal authorities would also be out on the same mission and, as the Federals took precedence over the locals, not to interfere with Augur's men. His next step was to send a message to Augur, explaining what his men were doing, and assuring the general that any witnesses at police headquarters who had any pertinent information would be sent on, at once, to Augur himself. There is nothing to show that Richards told Augur that he recognized the actor, Booth, as the assassin, but, on the other hand, it is not possible that Richards would have withheld the identity of the self-advertised killer of the century. Then too, the name of Booth could hardly have been news to Augur, because that was the only name being bandied on the streets.

At Rullman's Hotel, 456 Pennsylvania Avenue, the gaiety at the bar was silenced. Bartender John R. Giles had just announced the news. Most of the drinkers left at once. Over in a corner, at a table, Mike O'Laughlin and his three companions from Baltimore shook their heads in a drunken daze. Three of them could not believe the news. Mike could. And he could even guess the name of the assassin.

Surgeon General Joseph K. Barnes examined the injured at the Seward home. He bound the secretary's wounds and said that there was "severe loss of blood, and shock. If the patient recovers from the shock, he will probably live." He examined Frederick Seward and found a "double fracture of the cranium, profuse bleeding, no pulse, inability to speak." Frederick, he thought, might die.

At Ford's Theatre, William T. Kent talked his way back into the President's Box. He told the officer in charge that he was the one who had given a penknife to Surgeon Leale and, when he got home, he found that he had lost his house key. He was searching the box when his foot kicked against something loose and he picked it up.

"I have found the pistol!" said Kent.

A man came into the box, introduced himself as Mr. Gobright of the Associated Press, and said that he would give it to the police. Mr. Kent gave him the gun.

A few minutes before this, the police had been clearing the theater of curiosity seekers, yelling "All out! All out!" when Isaac Jaquette hurried out of Box 7 and, in the corridor, tripped over the wooden bar which had been used to hold the white door closed. Jaquette got out of the theater with it, took it home to his boardinghouse, and pointed to the drying blood on the bar. A Union officer asked for a piece of the bar as a souvenir, and Jaquette got a saw and cut it off. The officer studied the piece of wood, looked at the blood, and said that he did not want it.

The blood was not Lincoln's. It was Major Rathbone's.

The major, who sat in the Petersen parlor with Mrs. Lincoln and Robert, and the Misses Harris and Laura Keene, suddenly fell unconscious from loss of blood and was taken home. For the rest of the night, Robert either sat with his mother, or stood behind the head of the bed looking down at his father's face. The narrow hall was heavy with the tramp of boots, inbound and out, and from out in the street the roars of the crowds could be heard and the cursing of cavalry officers who rode through the people trying to clear the street.

In the dimness of the parlor, Mrs. Lincoln sat staring at the ruddy coals in the grate across the room. She said little. Now and then, she looked for assurance to the two women who flanked her. But, when the assurance had been given and received, men walked in and gravely offered their condolences, as though the President was already dead. This led to wild outbursts of grief, and repeated requests to "take me inside to my husband." When she got in the small bedroom, she looked, screamed, and fainted. The Rev. Dr. Phineas Gurley, with muttonchop whiskers quivering, uttered words of encouragement which he did not feel. Each time that Mrs. Lincoln made the trip to the sickroom, the doctors were warned ahead of time and placed fresh napkins under the President's head. Once, she stood looking down at him, supported on both sides, and the tears had made dry furrows in her face powder.

"Live!" she cried. "You must live!" She turned to the doctors. "Bring Tad. He will speak to Tad. He loves him so."

Back in the parlor, she sat looking at the coals, and she spoke aloud to herself: "Why did he not shoot me instead of my husband? I have tried to be so careful of him, fearing something would happen, and his life seemed to be more precious now than ever." Her tone changed and she spoke imperiously: "I must go with him!" Then silence and a loud demand: "How can it be so!" Robert crouched before her, rubbing her hand in his and murmuring: "Mother, please put your trust in God and all will be well."

Dr. Robert King Stone, the Lincoln family physician, arrived. He was a tall man in a stovepipe hat and he removed his gloves and coat while looking at the President and listening to the doctors tell of their findings. He asked for a basin of warm water and washed his hands. Then, sitting on the bed facing Lincoln, he lifted the head with his left hand around the neck and stuck the small finger of his right hand into the bullet hole. He probed a moment, withdrew the finger, studied the eyes now bloodshot, and pinched the cheek lightly.

"This case," he said, "is hopeless. The President will die. There is no positive limit to the duration of his life; he is tenacious and he will resist." He arose and walked back to the basin. "But death will close the scene," he said.

Technically, Leale was still in charge of the case and, when Dr. Charles Sabin Taft returned to the bedroom and asked permission to give a mixture of brandy and water to the patient, Leale said no. It might induce strangulation. Doctors Stone and Barnes were in the back parlor, and Taft went to them to get an opinion. He came back and said that it was their opinion that it might help the patient.

"I will grant the request," said Leale, "if you will please at first try by pouring only a very small quantity into the President's mouth."

Taft measured off about a third of a teaspoonful and parted the patient's lips and watched the liquid run into the mouth. The President did not swallow. At once there was a laryngeal obstruction and partial suffocation. Leale pushed Dr. Taft aside, opened the mouth and pushed his hand down as far as it would go and pulled at the base of the tongue. In thirty seconds, breathing resumed.

The doctors held a conference in the bedroom and they agreed that Surgeon General Barnes should take charge of the case. All decisions would be in his hands. Young Doctor Leale, dark, handsome, still youthful enough to have more heart than head, decided to sit with Mr. Lincoln until the end, and to render whatever help he might. As the doctors talked, Leale wrote a note: "Left side of patient's face begins to twitch. The mouth is pulled sharply to left in a jeer. After 15 minutes, it stops."

Another young and bright man, Corporal Tanner, reached home to find that his room, and his tiny balcony, situated next door to the Petersen house, were jammed with boarders. He had no privacy in his quarters, and no one seemed disposed to leave. So he accepted the situation with grace, and fought his way out on his balcony to look at the roaring black mass of humanity below.

He had just achieved a front-row view when the crowd hushed. Next door, General Augur had come out on the brownstone stoop and was waving his hands for quiet.

"Is there someone," the general hollered, "who knows shorthand?"

Albert Daggett, a postal card contractor, was standing beside Tanner and he cupped his hands and yelled down: "There is. He lives here."

The general looked up, and said: "Well, then. Send him down."

Tanner fought his way off the porch and shouldered his way through his room. He picked up two pencils and a fat pad, put on his uniform cap, and squeezed through the mob outdoors with the help of soldiers. At the Petersen home, he was conducted to the rear parlor, passing the patient en route. He recognized Mr. Stanton and Chief Justice Cartter, both sitting at a small library table. General Augur said that this soldier could take shorthand, and Stanton nodded and Tanner took a seat at a small round marble-topped table. Augur told Tanner that the other people in the room were witnesses, and that more were in the hall, but that the taking of their testimony in longhand had proved impossible.

The man they wanted was riding his mare up Good Hope Hill on the Maryland shore. At the top of the hill, Mr. Polk Gardner, coming into Washington City and not at all sure that he would be permitted to cross the bridge, stopped his horse. In the clear, low moonlight, Gardner saw a rider coming uphill. What interested him was that the rider was trying to prod the horse into running up the grade, and the horse would race a hundred feet, and lapse into a walk. The rider spurred the animal again. And again.

When he came abreast of Mr. Gardner, he stopped and said: "Good evening. Can you tell me if a horseman passed ahead of me?"

Gardner said no, that he had been on the road an hour, and had seen no one riding away from Washington City.

"Does not the road to Marlboro turn to the right a ways down here?"

"No," said Gardner. "You keep to the straight road."

"Thank you." Booth pounded off into the darkness. He now knew that neither Paine, nor Atzerodt, nor Herold had got out of Washington before him. He also knew that, if he was being pursued, any information Mr. Gardner might give would send the Federal patrols running off in the direction of Marlboro, instead of toward Surrattsville.

Polk Gardner barely got down Good Hope Hill when he saw a second rider. A teamster with a load of vegetables for Washington drove a little ahead of Gardner and the second rider stopped and spoke to this man. Gardner did not hear the words but, in a moment, the second rider was off, less than half a mile behind the first.

Booth pressed on until he heard hoofbeats behind him. He moved the mare into a stand of trees and waited. The rider went by and Booth came out yelling "Halt!" It was David Herold, and for the rest of the ride into Surrattsville the two had a lot to tell each other. The actor was certain that he had killed Lincoln instantaneously, with the laughter of hundreds ringing in his ears.

He was not so certain about his leg—it was either a bad sprain or a break. He hoped that it was a sprain. Herold said that he was sure that Paine had killed the Secretary of State because he was waiting below when a colored boy came running out yelling murder, and, with Augur's men only a few doors away, Herold couldn't afford to wait any longer. If the conspirators had lost any men, they would be Paine and Atzerodt.

The chief conspirator was now in bad and steady pain and he asked Herold to switch mounts with him. The mare had a bouncy walk and a rocking-horse trot. They switched, and continued on their way. The road was straight now and it fell away into a small valley. At the bottom, it was chill enough to see the horses' breaths. They talked of the guns and whiskey and Surrattsville and they talked of using a ferry north of Port Tobacco in case Atzerodt did not escape alive. But most of all, they talked of how they had crippled the North, when all was almost lost. Booth regarded himself as a modest hero, one who would never be boastful but one who, at the same time, expected that every loyal Southerner would be eternally grateful to him. In this, too, he was sincere.

For the next eight hours, the United States was run by a dictator. In the back parlor of Petersen House, and, at times, in the sitting room behind the front parlor, Edwin McMasters Stanton sat with the country under his thumb. And he had the dictator's gift for quick, and sometimes erroneous, decisions. In all, he did as well as anyone could have—perhaps better.

He convened a special court of inquiry, with Justice Cartter administering the oath to witnesses, and Stanton doing the questioning and Corporal James Tanner taking the testimony. Generals and senior officers were his messengers. The Cabinet members either took orders or remained silent. The steady stream of witnesses to the crime almost matched the steady stream of soldiers coming and going. When Stanton moved, he moved fast.

He ordered guards placed around the homes of all Cabinet members and ranking officials. He ordered the confiscation of Ford's Theatre and the arrest of "every human being" in its service. He sent an officer to William Dixon, Chief Engineer of the Washington Fire Brigade, ordering that all engines and apparatus be kept in a state of readiness because Stanton expected mass arson after mass killing.

Mr. Stanton announced to General Augur that this plot had a broad base—that the actual assassins were hirelings of the Confederacy—and that hundreds of terrorists were in Washington City this night. He wanted 150 policemen, 500 military policemen, the United States Secret Service, the spies of the Bureau of Military Justice, and the 8,000 soldiers in encampments in and around Washington to be ordered out at once to seek out and arrest these terrorists. It did not occur to Stanton or Augur that these soldiers, at large with visions of rapid promotion and high cash rewards, could open a reign of terror in the city.

The Secretary of War, his vest open, peering over the tops of his glasses, sat in a corner of the room farthest from the door. When he used the sitting room near the front of the house, the folding doors between it and the front parlor were pulled together, but the angry voices of Stanton and Cartter could be heard by Mrs. Lincoln and by Robert. Two soldiers stood outside the door with bayoneted rifles.

The witnesses were terrified and many who had come into the dark hall sure of the facts answered whisperingly that they did not remember. Top-hatted statesmen brushed by the witnesses en route to the little bedroom, or en route home. Generals and admirals joined the shuffling queue. Stanton was busy and Stanton was impatient. He fired questions, listened to stammering answers, asked more questions, wrote telegrams, denied requests, ordered arrests, paced the floor, stroked his perfumed beard and brusquely ordered citizens to leave at once.

Some, in good faith, gave poor answers. Lieutenant Crawford, who had sat on the aisle in Row D with Captain Theodore McGowan, said that a man passed them twenty minutes before "this" occurred. McGowan said: "Sir, I remember that a man passed me and inquired of one sitting near who the President's messenger was, and learning, exhibited to him an envelope, apparently official, having a printed heading and superinscribed in a bold hand. I could not read the address and did not try. I think now it was meant for Lieutenant General Grant. The man went away."

And Clara Harris, weeping, said: "Nearly one hour before the commission of the deed the assassin came to the door of the box and looked in to take a survey of the position of its occupants. It was supposed at the time that it was either a mistake or the exercise of an impertinent curiosity. The circumstances attracted no particular attention at the time. Upon his entering the box again, Major Rathbone arose and asked the intruder his business. He rushed past the major without making a reply, and fired. . . ."

Like scores of others, Harry Hawk said: "I believe to the best of my knowledge that it was John Wilkes Booth." Then, like many others, he was shaken by the enormity of the crime and he said: "Still, I am not positive that it was him. I only had one glance at him as he was rushing towards me with a dagger and I turned and ran and after I ran up a flight of stairs, I turned and exclaimed: 'My God! That's John Booth!' In my own mind, I do not have any doubt but that it was Booth."

Harry Phillips, who was to have sung "Honor to Our Soldiers" in the late intermission, was outside his dressing room when he heard the shot. He told Mr. Stanton and Justice Cartter that he ran downstairs in his shirtsleeves, heard actors and stagehands saying that Mr. Lincoln had been shot and that it was John Wilkes Booth who had shot him. "Are you certain it was Wilkes Booth?" he had asked Harry Hawk.

And Hawk had said to him: "I could say it if I was on my deathbed."

The parade of witnesses all said "Booth." Ferguson, who took the little girl home, stopped at the D Street precinct to tell the police that he saw Booth go into the box, heard the shot, and saw Booth leap out of the box. Then he took his place in line to tell Stanton and Cartter and Augur the same thing.

"In fifteen minutes," said Corporal Tanner, "I had testimony enough to hang Wilkes Booth higher than ever Haman hung."

At police headquarters, Superintendent Richards interrogated seventeen witnesses. The blotter noted:

At this hour the melancholy intelligence of the assassination of Mr. Lincoln, President of the United States, at Ford's Theatre was brought to this office, and the information obtained from the following persons goes to show that the assassin is a man named John Wilkes Booth:

E. D. Wray, Surgeon General's Office
J. S. Knox, 25 Indiana Avenue
Joseph B. Stewart, 349 K Street
Capt. G. S. Shaw, General Augur's Staff
C. W. Gilbert, 92 and 94 Louisiana Avenue
James B. Cutler, New Jersey Avenue
Jacob G. Larner, 441 F Street
James Maddox, Ford's Theatre
Anthony Lully, 406 K St. between 9 and 10
W. S. Burch, 333 F Street
John Deveny (an ex-Army officer)
Harry Hawk (the actor who was on the stage when the assassination occurred)
John Fletcher, 299 E. St. (Naylor's stables)

Andrew C. Mainwaring, Soldier's Home
William Brown—
John Gratton, Record Hospital
J. L. Deboney—Ford's Theatre—boards next door to
Callan's Drug Store

No word went out from Petersen House to apprehend Booth. Many fine literary minds have read into this an assent to the assassination on the part of War Secretary Stanton. There is no evidence in any record to support this. The opposite is true. Stanton felt stronger than the President and felt protective toward his person. He had seen Lincoln grow from a despised, unwanted figure to that of an almost sanctified statesman. Now, when his fears had been realized, when he saw the results of withholding the services of a man like Eckert through a lie, when he stood beside two beds and looked down at Seward and at Lincoln, his mind refused to accept the fact that the majesty of the United States Government had been affronted by one man, an actor. To him, Booth was small game. Stanton's function, as he saw it, was to stop the *pending* assassinations rather than to apprehend the perpetrators of the Lincoln shooting.

In support of his stand of a widespread plot, the commercial telegraph system went dead at 10:30 P.M. that night. Major Eckert and his chief visualized this as isolating Washington City from the rest of the North. To achieve it, both agreed that a large band of men must be working together because many wires would have to be cut simultaneously. At the same time, Eckert decided that his U.S. Army wires must have been tapped, and so, when Stanton wanted to send a message to a garrison in the District of Columbia, Eckert dispatched it all the way to Old Point, Virginia, with a request that it be relayed back to the proper garrison in Washington. In effect, Mr. Stanton was at pains to outwit men who did not exist.

At 11:45 P.M. the first order to apprehend went out from Augur. This was ninety minutes after the shooting, and Booth's name was not mentioned.

Colonel Nichols:
I have sent to arrest all persons attempting to leave the city by all approaches. Have telegraphed to troops on the upper Potomac to arrest all suspicious persons—also to Gnl. Slough at Alexandria and Gnl. Morris at Baltimore—All our own police and detectives are out. No clew has yet been found by which I can judge what further steps to take. Can you suggest any?

Respectfully,
C. C. Augur.

Colonel Thompson, at Darnestown, was ordered by Stanton to have his men patrol the area north of Washington City. THE ASSASSINS ARE SUPPOSED TO HAVE ESCAPED TOWARD MARYLAND, he wired. The secretary meant Montgomery and Anne Arundel counties to the north and east. Although Augur was beside him most of the night, Stanton sent a message to him too, ordering that no person suspicious or unknown be permitted to leave Washington City this night.

Just before midnight, Mr. Stanton further dissipated the Federal efforts when he sent another message to General Slough in Alexandria: IT IS NOT KNOWN IN WHICH DIRECTION THE ASSASSIN HAS ESCAPED.

At Fairfax Courthouse, General Gamble had eight hundred men on the roads between his headquarters and Leesburg. A message, directed to him via Major Waite, said: ORDER GENERAL GAMBLE TO ALLOW NO ONE TO PASS HIS LINES. TO ARREST EVERYONE WHO ATTEMPTS. STANTON.

Alarms went out to Winchester, Harper's Ferry, Cumberland, Baltimore, Annapolis, Acquia Creek, Relay House, almost everywhere except on the fat foot of Maryland which lies between the Potomac and the Patuxent. And that's where Booth was.

The hysteria was contagious and the generals and colonels were apprehending and arresting without reason. General Morris wired Stanton from Baltimore: THE MOST VIGOROUS MEASURES WILL BE TAKEN. EVERY AVENUE IS GUARDED. NO TRAINS OR BOATS WILL BE PERMITTED TO LEAVE THIS DEPARTMENT FOR THE PRESENT. In reply, Stanton had to order the general to permit three trains, laden with food for Washington City, to proceed out of Baltimore.

Perhaps the first official word regarding Booth was contained in a message written in longhand by General Augur and sent to Colonel Gile, Commander, Reserve Corps, Washington:

> The Major-General commanding directs that you
> detail a commissioned officer and ten enlisted men to
> accompany train which leaves this city for Baltimore
> April 15. Shortly after leaving the city, the officer in
> charge will search every car in the train and arrest, if
> found, J. Wilkes Booth and other parties whom you
> may deem it for the interest of the service to appre-
> hend. At each stopping place or station this search
> will be made. The party will in each case return to
> Washington by the train leaving Baltimore first after
> its arrival there, and carry out the same instructions
> on the return trip.

Mr. Stanton persuaded Mr. Welles to issue like instructions to the navy. Steamers were ordered to patrol up and down the Potomac, looking for fugitives close to shore. Down at the mouth of the river, Commander Parker at Saint Inigoes was ordered to bottle up the river. Point Lookout was ordered to stop any and all vessels proceeding south on the Potomac and to hold all persons aboard until further orders.

Every avenue of escape out of Washington was closed except the Navy Yard Bridge leading to southern Maryland, and no orders were issued about that one because, for a long time past, it had been closed every night at 9, and the attacks did not occur until 10:15. No one could have left the city that way.

At 11:55 P.M. Sergeant Silas Cobb, in charge of the bridge detail, was surprised to find that his night relief man was early. They swapped news and the relief sergeant said that he had nothing to tell because, until a few minutes ago, he had been sleeping in barracks. Cobb said that matters were quiet on the bridge. A few Marylanders, after a night of fun in Washington City, had been permitted to cross. That was all.

Sergeant Cobb went to bed.

At the commercial telegraph office, Mr. Dwight Hess, manager of Grover's Theatre, left a telegram to be sent to Mr. Grover in New York when service was resumed:

PRESIDENT LINCOLN SHOT TONIGHT IN FORD'S
THEATRE. THANK GOD IT WASN'T OURS.

The Final Hours

12 midnight

Among the small figures of history, John Fletcher stands as a man of tenacity. At this hour, he was still trying to find a horse. He went to the police station on Tenth Street to register a complaint against David Herold and, in the light of his own exasperating problem, could not understand why, at midnight, the precinct station was so full of fashionably attired ladies and gentlemen. He knew that the President had been shot, but Fletcher had given no further thought to it.

Now he spoke to Detective Charles Stone and asked if a roan horse had been picked up. Stone said: "Who are you?" Fletcher told him, and now, in a flash of revelation, the stable foreman began to wonder why Atzerodt wanted his horse made ready as late as 10 P.M. and why Herold had panicked and run off twenty-five minutes later when Fletcher asked him to get off the horse.

Stone said that a horse had been picked up, and that he would accompany Fletcher down to General Augur's headquarters, where they were trying to identify it.

Augur had returned from another conference with Stanton, and sat at his desk. Witnesses were waiting in an anteroom. Stone said that the stableman had rented a roan and the general asked if Fletcher remembered the name of the renter.

"I do," said Fletcher. "His name is Herold and he is young, not more than twenty-one or -two, he looks like a boy, and I can tell you that I followed him to the Navy Yard Bridge before I lost him."

There Augur had the key to the conspiracy. But he shook his head sadly. The description did not fit John Wilkes Booth, and Booth was the man that Augur wanted. Besides, the general agreed with Mr. Stanton that the attempts on the lives of Seward and Lincoln were probably executed by the same man. There was no room in Augur's mind for a boyish-looking assassin.

The general nodded toward a flat saddle and bridle lying on a chair.

"Do you know anything about that?" he said.

Fletcher, an expert in such matters, walked over and fingered the leather. He turned the bridle over in his hand.

"I do," he said.

"What kind of a horse had that saddle and bridle on?"

"A big brown horse, blind of one eye; a heavy horse with a heavy tail; a kind of pacing horse."

Augur was now interested. An hour ago, a big brown blind horse had been picked up in East Capitol near Fifteenth Street in a farm district. He did not know whether this horse figured in the assassinations, and his best thinking was that there was no connection. He could not imagine desperate killers on brewery horses.

"Who rode this horse?"

Fletcher said he could not remember the name, but he could get the name from his stable records. He had stabled the horse that wore these trappings. And he knew the little man with the funny name. But he pointed out that this man had used the horse until *recently*. Who had him tonight Fletcher had no idea. The detective accompanied him back to the stable and the Irishman returned with the name of George Atzerodt. This meant nothing to Augur. He was even less interested when Fletcher said that Atzerodt claimed that he had sold the horse and equipment in southern Maryland.

Now the general had the names Booth—Herold—Atzerodt.

He needed only the name of Lewis Paine, who had ridden and abandoned the blind horse tonight, to complete the roster of all the conspirators. Augur felt that he was wasting too much time with the stableman. He had important eyewitnesses waiting outside. He excused Fletcher. Had he chatted with the irritated Irishman a little longer, he would have learned that Atzerodt and Herold had a friend named Booth, and that Booth was accustomed to lending his horses to a man on H Street named Surratt, and that, quite often, these men rode to Surrattsville. Fletcher even knew that John Surratt's father had owned a tavern in Surrattsville.

A horse car lit up the East Capitol section as it moved slowly down Sixth Street to the Navy Yard. Some of the night-shift men were on their way to work. A few late revelers dozed in the seats. The car stopped at A Street and George Atzerodt got aboard. Most of the men in the car saw him as an apologetic little man in a round hat, a man who excused himself as he pushed toward the back.

Atzerodt had to find his friend who owned the store. He was drunk and apprehensive, and he needed rest. He would sleep on the floor of the store. He was still pushing toward the back, and the horse in front was walking slowly between the undulating rails when someone tapped him on the back. It was Washington Briscoe, the man he was looking for.

"Have you heard the news?" said Briscoe.

"Yes," said Atzerodt. He asked Briscoe if he could have permission to sleep in front of the store.

"No," said Briscoe. "I cannot do that, George."

"I will make no noise," said Atzerodt.

"I am very sorry, George. The owner is at my place and I won't have any guests tonight."

"As a favor."

"No."

Both got off at the Navy Yard. George Atzerodt asked once more. The answer was no. The conspirator looked sad.

"I will go back to the Kimmel House," he said. This was a neighborhood nickname for Pennsylvania House, a four-and-five-men-to-a-room hotel on C Street near Fourth. He waited for the horse car to turn around. George Atzerodt, right now, was within a few streets of the Navy Yard Bridge, but he had no desire to join the other conspirators.

It was a night of frustrations and heartache. In Augur's quarters, a young captain walked in, saluted, and asked for permission to lead a mounted squad to the conspirators and capture them tonight. The general had little sympathy with inspirational heroes.

"No," he said. "Permission is denied. Instead, you will remain here, Captain, for whatever emergency duty may be assigned to you."

The captain remained. All night long at headquarters, he waited for someone to give him something to do, but Augur's staff did not know him and did not need him. Thus another opportunity was lost, for this was Captain D. H. L. Gleason, the man who worked with Louis Wiechman and who reported that a plot was being devised against the President at a boardinghouse called Surratt's, 541 H Street. Gleason knew that Booth and the other conspirators sometimes met at a tavern in Surrattsville, and that one or more of them came from Port Tobacco.

Augur wouldn't give him the men to make the capture. And Gleason volunteered no information.

At 12:30 A.M. telegraph service was restored. The wires, it was learned, had not been cut. Two wires in the main battery had been crossed and all service had been shorted. Government officials tried to make this look like a planned Confederate move, but the Washington City manager of the office was a prosaic businessman who insisted that wires are sometimes crossed and, when they are, service is temporarily suspended until they are located.

It was a few minutes past midnight when Booth and Herold passed the crossroads at Surrattsville. They were now more than eleven miles south of Ford's Theatre in rich farm country. The air was chill, the moon bright, as they looked at the tavern on the left. It sits back one hundred or more feet from the road on a raised biscuit of land. A long porch, with rockers, spread across the face of the building. At the far right a small farm road was cut into the land and led up to a drafty barn and a well pump. A single light burned in the barroom.

The two swung up the dirt road and stopped abreast of the porch. Herold dismounted. Booth did not talk. He was in pain and he shifted this way and that in the saddle to try for small comfort. The horses panted and foraged in the dead grass as Davey bounded up on the porch and went inside.

Mr. Lloyd was on a couch, sleeping. The bar was empty and the bartender had gone home.

"Mr. Lloyd," Herold said, shaking him, "for God's sake make haste and get those things."

Lloyd sat up slowly. He looked at Herold. He was drunk. "All right," he said. "All right." He shuffled off upstairs and got two carbines, the field glasses, one cartridge box and he stopped behind the bar and drew a quart of whiskey. He did not bring the rope or the monkey wrench.

Outside, Herold said: "He has no brandy. Only whiskey."

John Wilkes Booth drank deeply, looked around at the silvery farmland, took another, and gave the bottle to Herold. Booth took the field glasses and ordered Herold to leave the carbines.

"It's a good gun," Herold said. "We ought to have something."

"Government issue," said Booth. "They will slow us up."

Lloyd was wavering on the lawn and, by standing with his legs spread, kept from falling.

"Is there a doctor in this country?" Booth said. "I have broken my leg."

Lloyd looked up at the rider. "Doc Hoxton is down the road," he said. "About a half a mile, I guess. But he don't practice. Told me so himself."

Booth nodded to Herold, who took the equipment back into the tavern but kept one of the carbines and some of the cartridges. Davey got on his horse.

"We killed the President," Booth said, "and Seward."

If Booth looked for applause, or incredulousness, or awe, there was none of it. Lloyd kept looking up at him stupidly.

"Don't you want to hear the news?" the assassin said.

"Use your own pleasure about that," Lloyd mumbled.

Herold handed Lloyd a silver dollar. The horses swung and were gone.

The pain forced Booth to change his plans. Instead of riding eighteen miles straight south to Port Tobacco, and escaping across the broad bend of the Potomac to the state of Virginia, he had to find medical assistance. The leg would not wait. The only doctor he knew in the whole area was Dr. Samuel Mudd, the humorless farmer at Bryantown.

Booth did not trust him, but he had to find a doctor. Mudd's place was about seventeen miles from the tavern, down past Waldorf and then to the southeast. It would have to be done before daylight, and that meant that a lot of hard riding lay ahead. Because he did not trust Mudd, John Wilkes Booth determined to wear the whisker disguise he had brought along, and the big muffler. He told Herold that they would say that the horse had stumbled on the road and Booth had been pitched off and had broken his leg. Once the leg was treated, they would ask for cross-country farm roads to take them to

Port Tobacco. The delay should not cost them more than an hour.

They headed down through T.B. and kept the horses moving.

In Washington, Major James O'Beirne of the District Board of Enrollment had heard the name of Booth so many times that he asked the theater people where Mr. Booth lived. They told him the National Hotel, at Sixth and the Avenue. The major called Detective William Eaton and ordered him to go there at once, find Booth's room, and take charge of it and everything in it in the name of the War Department of the United States.

In forty-five minutes, Eaton was back with the news that no one at the hotel had seen the actor since early last evening. However, the detective had brought back with him a trunk, a lot of papers, some letters and effects. O'Beirne ordered these turned over to Lieutenant William H. Terry to assess. The only item of importance turned up was a letter from one Samuel Arnold begging Booth to desist from a complicated plot.

It was well past midnight when Stanton ordered his enforcement subordinates to Petersen House for a conference. Major Richards of the Washington police attended. So did General Augur for the military and Major O'Beirne of the United States Marshal's office. Chief Justice Cartter was there, and Assistant Secretary of War Charles Dana arrived, and remained to write dispatches for Stanton. Captain William Williams, who had invited John Wilkes Booth to have a drink with him, was also there.

Corporal Tanner was ordered not to take any notes. Decisions good and bad were made at this meeting. Stanton announced to those present that he had indisputable proof that Booth fired the shot from outside the box door. The proof was that an investigating officer had located the bullet hole in a panel of the door. It was pretty definite, he admitted, that the assassin was John Wilkes Booth, but he wanted no public announcement of this yet. Stanton gave no reason for not announcing at once the shooting of the President, but those around him felt that he expected momentarily to arrest the assassins and that he wanted to announce the shooting and arrest at the same time.

He directed General Thomas M. Vincent to take charge of Petersen House and to be responsible for those persons who were admitted. He asked Dana to send a wire to General Grant, at the Philadelphia railroad terminal, telling him that Lincoln had been shot, and for Dana to contact the president of the Baltimore & Ohio Railroad and have a special train for Grant ready to leave Philadelphia for Washington. By telegraph, he asked Chief Kennedy of the New York police for some "good detectives." He wrote a terse note to Chief Justice Salmon Chase that Lincoln was dying and to be ready to administer the oath of office to Vice President Johnson. He notified Johnson that Lincoln was dying, and sent one of O'Beirne's men to protect him.

John Lee was the man sent to Johnson's hotel.* Like Richards, he was a policeman's policeman. When he arrived at Kirkwood House, he made a perfunctory stop at the Vice President's room, then called the assistant manager and asked to be taken to the roof. Lee examined the building from top to bottom. He found that the roof abutted other roofs and the building could be entered by a skylight. He also found a backyard fire escape which would permit any prowler to get into the hotel by the second floor.

Lee sent for another man to guard the upper hotel levels and he went into the bar for a drink. A customer asked Lee if he was a policeman and, on being assured that he was, said that there had been a suspicious-looking man who had taken a

---

* Lee swore at the trial of the conspirators that this happened on April 15. He probably meant after midnight of the fourteenth.

room yesterday and who had been asking questions about the Vice President. Mr. Lee asked the night manager for the hotel registry and asked him to point out the names of persons he could not vouch for as regular customers. The man pointed to the name G. Atzerodt.

The detective asked to see Mr. Atzerodt and was told that he wasn't in. Lee said that he would like to see Mr. Atzerodt's room, and insisted that the manager accompany him. They went up and knocked on the door twice. Lee called out the name Atzerodt. He said he wanted a key to the room and the manager, embarrassed, said that Atzerodt had the only one.

"I do not like the appearance of things," said Lee, "and I must get into this room."

He went back downstairs and asked permission of the proprietor to break down the door. The owner said that if the matter was important, all right. Lee said it was. He went back upstairs and, with the assistance of the night manager, broke the door.

"Stand in the doorway," Lee said. The detective lit the gas light and searched the room slowly and carefully. He took a black coat off a door and laid it on the iron bedstead. He was searching the pockets when he stopped, slipped his hand under a pillow, and came up with a huge pistol.

"Now," he said, "I will have to send for Major O'Beirne. You stand right where you are."

Lee went downstairs and almost bumped into O'Beirne in the lobby. The major had just delivered Stanton's message to Johnson that the President was dying and to hold himself in readiness to take the oath as President of the United States. The detective told his boss what he was doing and what he had found and O'Beirne ordered him to continue the search and to report later.

Back in the room, Mr. Lee continued searching the black coat and found an Ontario, Canada, bankbook made out to

J. Wilkes Booth in the sum of $455. A large-scale map of the state of Virginia was found. A white handkerchief came out of a pocket. Along one edge was stenciled "Mary R. E. Booth." A second handkerchief was marked "F. M. Nelson." A third was marked "H." There was an empty envelope with the frank of Congressman John Conness on it, and a new pair of gauntlet gloves.

In a bureau drawer were three boxes of Colt pistol cartridges. Also a stick of licorice, a toothbrush and an unmarked colored handkerchief. There was a single spur on the dresser, a pair of socks and two collars, one size sixteen, one size seventeen.

Lee then turned up the carpet, section by section, and examined the floor underneath for saw marks. He examined the washstand and basin, the back of the chest of drawers and, when he reached the little wood stove in the corner, he squatted and sifted the cold ashes. Then he took the bedclothes, piece by piece, and studied them and ran them through his fingers. He tore the pillow open and felt inside. Between the bottom sheet and the mattress he made his last find: a large bowie knife.

The Secretary of War now had an unmistakable cross reference between John Wilkes Booth and George Atzerodt. And, from Fletcher through General Augur, he had a cross reference between Atzerodt and David Herold. At 12:50 A.M., the only person he knew nothing about was the Seward assassin, Lewis Paine. Two hours and thirty-five minutes after the attacks, Mr. Stanton knew who was wanted.

1 a.m.

The local train carrying General and Mrs. Grant chuffed into Philadelphia terminal an hour late. Except for a few officials, the stationmaster and some police officers, the train shed was empty. There had been no public announcements that the Grants were coming to Philadelphia and only the barest telegraph warning had gone to railroad executives.

The Grants were greeted effusively by the few. Both were tired and gratefully boarded a military ambulance, which took them down to the Camden ferry. They were waiting in the ferry house with their luggage when Stanton's telegram arrived telling of the assassinations in Washington. At the end of the telegram, the Secretary of War said that arrangements were being made to bring the general back to Washington by special train.

The general was shocked at the news. After some thought, he decided that he would accompany his wife to the Camden side of the Delaware, and then would return to Washington at once. However, on the ferry ride, he must have changed his mind because Grant went all the way to Burlington, New Jersey, with his wife, saw his two children, and then agreed to go to Washington.

Back at Petersen House, Stanton did not need the services of Grant. Stanton needed no one, in fact, and barely consulted the other members of the Cabinet. And yet there was no exultance in the power he wielded on this night. He assumed that only he could be trusted to keep his head in an emergency; only he could fathom the complicated moves which must be made, and only he could execute them with dispatch. On this night, only he issued orders, wrote messages, barked questions, threatened witnesses, summoned high personages, detained, arrested, disposed and took the reins of government as though all his life had been a training ground for this one event.

He was in this little sitting room not to weep, not to brood over a man he had often belittled, but in cold fury to play the part of the master policeman. He did not hesitate to issue orders even where he lacked power. He had no jurisdiction over the metropolitan police force and yet he ordered the day men to get dressed and patrol the streets. He held the news of all that was happening in his fist, and he refused to open it until he was ready. He it was who ordered that no news of the assassination be permitted in any of the military districts of the South. It would be days before Atlanta and Savannah and Mobile knew that Lincoln had been shot.

Edwin McMasters Stanton was boss.

One of the few poignant mistakes he made was when he ordered Attorney General Speed to draw up a formal note to the Vice President advising him that President Lincoln had died and asking him to prepare to assume the presidency at once. When it was completed, Stanton read it aloud and ordered General Vincent to "make a fair copy of it for the files." He heard a scream, and turned to see Mrs. Lincoln standing, her hands clasped in entreaty. "Is he dead?" she shrieked. "Oh, is he dead?"

The Secretary of War tried to explain that he was merely preparing for a grave eventuality, but Mrs. Lincoln was moaning and not listening. She was led back to the front parlor.

In moments of absolute quiet, the President's breathing could be heard in the several rooms on the ground floor of the house. Dr. Barnes noted that spasmodic contractions of both forearms had begun. The muscles of the chest became fixed and the patient began to hold his breath in spasms, emitting it in gusty explosions.

Senator Sumner, sitting near the head of the bed, took the President's left hand in his and, bowing his head to the bed, began to sob. Seeing this, Robert Lincoln began to weep. Dr. Charles Taft, leaving, said: "It's the saddest death scene I've ever witnessed."

Among those for whom this ordeal was difficult was Dr. Leale. He had a professional interest, and a personal interest. No one knew—and Leale did not mention it this night—that the young doctor had idolized Lincoln for a long time. On Tuesday, he had finished his surgical duties at the Soldiers' Hospital early so that he could stand in a crowd and hear Lincoln speak at the White House. He had come to stand in front of Ford's Theatre tonight, not to gawk, but to look upon the face of a man he loved. He had bought postcard pictures of Lincoln to hang in his room. To Leale, the sixteenth President was the greatest.

Now he was a doctor on a case. And his opinions were crisply professional as he worked through the final hours. A clot formed in the bullet hole every few minutes, and Leale insisted that he would remove them, and no one else. He remained at the President's side and sometimes, if the Surgeon General watched, he would see Leale holding the President's hand. He wasn't taking a pulse. He was holding the hand.

Dr. Leale had a reason for this. He thought that, just before death, reason and recognition often return to a patient for a brief moment. Leale held Lincoln's hand, as he explained later, so that if reason did come for a moment, "he would know, in his blindness, that he was in touch with humanity and had a friend."

In the back parlor, Attorney General Speed made the finished copy of the formal notification of Lincoln's death at 1:30

A.M. and called upon the members of the Cabinet to sign it. This was done at once. At almost the same time, Stanton decided to release the news, through General John Adams Dix, Commandant, New York. General Dix resembled Secretary Seward in looks, having the same gray hair, smooth cheek, and patrician air.

> *War Department April 15, 1865*
> *1:30 a.m.*
> *Sent 2:15 a.m.*
> Major General Dix,
> New York:
>
> Last evening, about 10:30 P.M., at Ford's Theatre, the President, while sitting in his private box with Mrs. Lincoln, Miss Harris, and Major Rathbone, was shot by an assassin, who suddenly entered the box and approached behind the President. The assassin then leaped upon the stage, brandishing a large dagger or knife, and made his escape in the rear of the theater. The pistol-ball entered the back of the President's head, and penetrated nearly through the head. The wound is mortal. The President has been insensible ever since it was inflicted, and is now dying.
>
> About the same hour an assassin (whether the same or another) entered Mr. Seward's home, and, under pretense of having a prescription, was shown to the Secretary's sickchamber. The Secretary was in bed, a nurse and Miss Seward with him. The assassin immediately rushed to the bed, inflicted two or three stabs on the throat and two on the face. It is hoped the wounds may not be mortal; my apprehension is that they will prove fatal. The noise alarmed Mr. Frederick Seward, who was in an adjoining room, and hastened to the door of his father's room, where he met the as-

sassin, who inflicted upon him one or more dangerous wounds. The recovery of Frederick Seward is doubtful.

It is not probable that the President will live through the night. General Grant and wife were advertised to be at the theater this evening, but he started to Burlington at 6 o'clock this evening. At a cabinet meeting yesterday, at which General Grant was present, the subject of the state of the country and the prospects of speedy peace was discussed. The President was very cheerful and hopeful; spoke very kindly of General Lee and others of the Confederacy, and the establishment of government in Virginia. All the members of the cabinet except Mr. Seward are now in attendance upon the President. I have seen Mr. Seward, but he and Frederick were both unconscious.

Edwin M. Stanton
Secretary of War

Thus the first news of the assassination was designed for release in New York. The reader may quarrel with the small errors Mr. Stanton made in his report, but the big one is that he knew the name of the assassin but did not mention him. This, in effect, prevented the newspapers from broadcasting an alarm for Booth, because it left them with no high official source to attribute the name to. At this hour, every newspaper in Washington and the Associated Press in New York knew that the actor Booth was the man being hunted as the assassin, but, when they saw the Stanton statement in General Dix's hands, most of them decided not to use Booth's name; a few decided to hint by announcing that the "scion of a famous family of actors is being sought."

At military headquarters in Washington, General Augur had completed questioning most of the witnesses, and had time to think. The more he thought about stableman Fletcher's story of the blind horse, the more he thought that the conspirators might be traced through a stable. He ordered fresh patrols of cavalry out to canvass every stable in Washington City. If Booth rented a horse, he could hardly have concealed his identity. He wanted any stableman who had seen Booth brought to him at once.

The man arrived shortly before 2 A.M. He had seen Booth off on a small roan mare at four o'clock, he said. Augur listened. There was a sort of loose partnership in horses, the stableman said, between Wilkes Booth and a man named John Surratt. It was hard to tell which one owned the horses, because both men used them. Sometimes, either of them gave permission to a man named George Atzerodt and another man named David Herold to use the horses. And another strange thing: Booth had a big one-eyed gelding which only today he had ordered stabled in the back of Ford's Theatre.

It was then that some unremembered officer on Augur's staff figuratively snapped his fingers and, in substance, said: Just a moment. Didn't we get a report on an alleged kidnapping of President Lincoln some months ago, and wasn't someone named Surratt a part of that report? Didn't a woman named Surratt keep a boardinghouse somewhere nearby? Wouldn't that explain why, with the bridges closed and armed patrols all over the city, we have not been able to find John Wilkes Booth? Couldn't it be that he is hiding right now in Washington City?

It could.

2 a.m.

General Augur was still thinking about Booth and the board-inghouse when Richards, at police headquarters, made up his mind to raid the place. Before him he had the names of Booth, Herold, Surratt. The only one of the three with a known address was Surratt, and Richards remembered that the military had issued a confidential report on the Surratt house based on intelligence given by an informer in the War Department.

The superintendent of police called Detective John A. W. Clarvoe and explained his hunch. He asked that Clarvoe select a squad of good detectives and go up to H Street and raid the place now. Richards asked him to bring back Booth and Surratt, if they were there. Richards might have said: "Bring back anyone on the premises," but his hunch was still only a hunch and he did not want to be in the position of instigating terror raids at 2 A.M.

In fifteen minutes, Clarvoe was standing in front of the darkened house. H Street was deserted. He had ten men with him and he posted them carefully. One went to the back of the house, in the yard, one in the alley to the east, four men at the four corners of the building. Then Clarvoe, accompanied by Lieutenant Skippon, Detective Donaldson, Detective Mc-Devitt and Officer Maxwell, climbed the white stone steps to the parlor floor. Clarvoe pulled the bell.

The men could hear it jangle inside. Maxwell kept a hand on his gun. The detective was ready to pull the bell again when he heard a sleepy male voice from inside the door.

"Who is it?"

"We're police." The door opened slightly. "Who are you?"

"Louis Wiechman."

"Is John Surratt in?"

"No. He is not in the city."

"Does his mother live here?"

"Yes."

"I would like to see her."

"I'm sorry. She's in bed."

"It makes no difference. I must see her."

"All right. I will speak to her first."

The big boarder tried to close the door. Clarvoe and Skippon leaned on it and stepped inside. Wiechman was frightened. He stood with a small lamp, his nightshirt tucked into his trousers, his feet bare. He walked to the back of the house, Clarvoe a step behind him. The other policemen began to light lamps in the house. One went upstairs to the top floor, the other down to the basement dining room and kitchen.

Wiechman knocked on Mrs. Surratt's door and, through the closed panel, held a whispered conversation. Detective Clarvoe stepped closer and said: "Is this Mrs. Surratt?"

"Yes," said the woman behind the door.

"I want to see John."

"John is not in the city, sir."

"When did you see him last?"

"It must be two weeks ago."

Clarvoe signaled for another detective to take over the questioning of Mrs. Surratt. He turned to Wiechman and walked back to the sitting room.

"Do you belong here?" Clarvoe said.

"I do."

"Where is your room?"

Wiechman pointed.

"I want to see it," said the policeman.

Wiechman led the way upstairs, followed by Clarvoe and

Lieutenant Skippon. The boarder opened the door to his room and stepped inside. He turned the kerosene lamp up.

"Is that your trunk?" said Clarvoe.

"Yes sir."

As the policeman stooped to open it, Wiechman laid a timid hand on his shoulder. The boarder's eyes were pleading.

"Will you be kind enough to tell me the meaning of all this?"

Clarvoe straightened. "That is a pretty question for you to ask me. Where have you been tonight?"

"I have been here in the house."

"Were you here all evening?"

"No. I was down the country with Mrs. Surratt."

A second detective was rummaging through a closet. "Do you pretend to tell me that you do not know what happened this night?"

"I do," Wiechman said. "What happened?"

"I will tell you," Clarvoe said. From his pocket he pulled a piece of wilted collar and a small bow from a tie. The collar was stained orange. "Do you see this blood? This is Abraham Lincoln's blood. Wilkes Booth has murdered the President and John Surratt has assassinated Mr. Seward."

The boarder clapped a hand to his forehead dramatically. "Great God!" he moaned. "I see it all now!" He staggered. "Is it really true?"

Clarvoe rummaged through the trunk and Skippon, who had finished with the closet, nodded to him. They went downstairs with Wiechman. At the foot of the stairs, Mrs. Surratt was answering some questions asked by Detective McDevitt.

The boarder said: "What do you think, Mrs. Surratt? President Lincoln has been murdered by John Wilkes Booth and the Secretary of State has been assassinated."

Mrs. Surratt raised both hands above her head and said: "My God! You don't tell me so!"

Clarvoe closely watched the reactions of the landlady and the boarder to the news, and both seemed to him to be genuinely shocked.

"Mrs. Surratt," said the detective, "I am going to ask you a couple of questions and I want you to be very particular how you answer them because a great deal depends upon them. When did you see John Wilkes Booth?"

The little landlady thought for a moment. "Why," she said, "two o'clock this day."

"You mean yesterday?"

"Yes, yesterday."

"When did you last see your son John?"

"About two weeks ago."

"Where is he?"

"The last I heard, in Canada. I received a letter from him. There are a great many mothers who do not know where their sons are. What is the meaning of all this?"

Clarvoe started up the stairs again. He nodded to McDevitt: "Mack, you tell her." He went to the top floor and tried a doorknob. The door was locked. He heard a female voice say: "Who is it?" On the opposite side of the hall, a door opened and John Holahan came out in his nightshirt.

"John," said Clarvoe, "how do you do? What are you doing here?"

Holahan, half awake, squinted in the gaslight and roared: "And how are you, John? I board here. What's the matter?"

"How long have you been here tonight?"

"Why, I took a walk and got in early. Nine o'clock, I think."

"President Lincoln has been murdered."

Holahan reacted like the others. He took a staggered step backward.

"Great God Almighty!"

Clarvoe took the doorknob in his hand, and Holahan restrained him. "My little daughter is in there," he said.

Clarvoe tried to go into the other bedroom. "John," said Holahan, "my wife is in there. Let me talk to her and then you can come in."

Clarvoe waited outside of both doors. The boarder talked to Mrs. Holahan and said that she would be presentable in a minute. The detective wanted to know if there was anything upstairs, and Holahan told him a furnished attic, and led the way. Anna Surratt was there and, with her was the young girl who usually shared a bed with Mrs. Surratt, Honora Fitzpatrick.

"Do you mind," said Holahan, "if I warn them first that someone is coming in?"

"Go ahead."

Holahan stepped into the room, awakened the young ladies, and told them to get dressed, that police were in the house. The girls did not get dressed. In fright, they elected to cover their heads with bedclothes and remain where they were. Clarvoe searched the room and, realizing that he did not know who was under the bedclothes, said that he was sorry, but that he would have to see their faces. He pulled the bedclothes back a little and took a look.

He went back downstairs, met Mrs. Holahan, examined that bedroom, peeked into the one across the hall where young Miss Holahan was, and then joined the others down on the parlor floor. Clarvoe and Skippon went down to the basement and searched it. In the kitchen, they saw a Negro woman.

"Auntie," said Clarvoe, "is John Surratt in this house?"

The woman was badly frightened. She shook from cheek to heel. "Do you mean Mrs. Surratt's son?" she said.

"I do," Clarvoe snapped. "I didn't know she had a husband."

"I have not seen him for two weeks."

Clarvoe and Skippon went back upstairs and searched

that floor from front to back. They questioned the boarders in the parlor. They left. They took no prisoners.

Mr. Stanton would know about all of this in about an hour, but, at the moment, in Petersen House, he had lost his temper. Mrs. Lincoln had made one more trip to the deathbed. She was supported by Miss Harris and Miss Keene and she had leaned across the bed so that her cheek rested on her husband's. At that precise moment, the President expelled an explosive breath, and, as her ear was close to his mouth, the noise terrified her and she screamed and fell into a dead faint.

Stanton heard the commotion, the cries of the other ladies, and he came into the bedroom pointing a finger at the unconscious Mrs. Lincoln.

"Take that woman out," he said loudly, "and do not let her in again."

When the room had been cleared, Dr. Barnes ordered the patient to be turned toward the wall. The doctor sat on the bed and, with cotton soaked in alcohol, combed the black matted hair away from the round wound. Then, using a silver probe, he tried to locate and remove the bullet. The probe moved inward two inches, and met an obstruction.

Barnes asked his assistant for a long Nélaton probe. It had a tiny white porcelain bulb on the end. This, when inserted, passed the two-inch mark and continued onward diagonally across the brain. At a depth of four inches, it ran into an obstruction. Barnes turned the probe slowly so that segments of whatever the obstruction was would be found on the porcelain bulb. If it was a bullet, traces of lead would be found. He withdrew it. There was no indication of lead. The other doctors studied the probe and agreed that he had probably contacted a piece of loose bone which had been blown from the back of the skull by the impetus of the bullet. The Nélaton probe was tried again, without result. The Surgeon General, after a consultation, agreed that no further effort would be made to find the bullet.

Andrew Johnson felt that he had waited long enough to visit President Lincoln. Later, many would say that the Vice President did not want to go to Petersen House. Whether or not this was so, he received the Stanton message to be prepared to take the oath of office and at once insisted that he was going to Petersen House. Governor Farwell opposed it. He said that the future of the Republic was bound up with Johnson now, and that the Vice President should remain where he was. Major James O'Beirne was present, and he too opposed the visit.

Johnson said he would go anyway. At that, O'Beirne said he would summon a guard of soldiers. The Vice President refused. He wanted no guard, no carriage—he would walk. So Farwell and O'Beirne flanked the next President and walked him up Twelfth Street and across E to Tenth. Johnson said little. He had pulled his hat down hard over his eyes, raised his coat collar, and jammed his hands into his coat pockets.

Tenth Street was almost deserted. Cavalry horses were tied four and five to a picket post up and down the street and they looked dejected in the cool dampness of morning. A small group of civilians stood around Petersen House and two soldiers patrolled the front of the house.

The Vice President was shown into the bedroom. He stood with his hat in his hand, his hair mussed, looking down. He stood for a little while, never taking his eyes from the figure on the bed, not saying anything, not showing any emotion. Then he took Robert's hand and whispered a few words. He stopped in the back parlor and said something to Stanton, who looked up at him and nodded curtly. He went back through the hall, through the bedroom with the flickering jet, and into the front room. He took Mrs. Lincoln's hand in his and she looked up at him, whimpering.

Johnson walked back to Kirkwood House.

In New York and in Philadelphia and Chicago and Detroit and St. Louis and Boston, the morning newspapers were being made up. They knew. Now the editors were going to press with the biggest, saddest story of the age. Mourning rules were dropped into place by printers and many editors headed the story with the single big word "IMPORTANT!" This was followed by as many as fifteen and eighteen diminishing headlines which told, in brief, the facets of the story.

Radical Republican newspapers ripped out editorials which condemned Lincoln's "soft" peace toward the South, and in their place went brand-new editorials which mourned the loss of a great man. Cartoons which slandered Lincoln's features were tossed on the composing-room floor. Anti-Lincoln letters from irate readers were killed, the type distributed.

Mainly, the story ran down the left-hand column of page 1 and jumped from there to another page. Smaller sidebar stories, telling of the effect of the assassination on the national welfare, were run beside the main story. So were stories about Mr. Seward's assassination, and there were a few stories about Johnson and Stanton.

The editors were also exasperated. They told their readers about the greatest crime of the nineteenth century, but in the story there was no criminal. They jammed the reopened wires to Washington with questions. A few who had stories which mentioned Booth removed the name from the copy because it seemed libelous. On the wires, the editors asked for confirmation, by a high official, of Booth as the assassin. Associated Press members were confused because they were asked, at one time in the night, to "kill" the story.

The editor of the *National Intelligencer,* who did not know that Booth had tried to give him a news beat and a confession in a letter, sat down at 2 A.M. to write an editorial in longhand.

He gave his lead a lot of thought and then he wrote: "Rumors are so thick and contradictory that we rely entirely upon our reporters to advise the public of the details and result of this night of horrors. . . . We forbear to give the name of one of the supposed murderers, about whom great suspicion gathers. . . . At the Police Headquarters it is understood that Mr. Hawk, of Laura Keene's troupe, has been held to bail to testify to the identity of the suspected assassin of the President, whom he is said to have recognized as a person well known to him."

The Washington *Chronicle,* by comparison, stated the case for confusion as well as any newspaper: "We then as-certained that the police were on the track of the President's assassin, and found that a variety of evidences, all pointing one way, would in all probability justify the arrest of a char-acter well known throughout the cities of the United States. Evidence taken amid such excitement would, perhaps, not justify us in naming the suspected man, nor could it aid in his apprehension."

Almost alone among the big daily newspapers, the New York *Tribune* named names: "Laura Keene and the leader of the orchestra declare that they recognized him [the assassin] as J. Wilkes Booth, the actor."

Over on C Street, John Greenawalt had just retired at Pennsyl-vania House. He owned the hotel and the houseboys seldom disturbed him because, when he was sleepy, he was irritable. However, he had just become comfortable when a boy knocked, came in, and said: "There is a man came in with Atzerodt, and he wants to pay for a room."

The houseboy was wrong. Atzerodt had not come in with anyone. He was in the lobby, dozing on a settee. He wanted a place to sleep, but he had no money. When a paying customer had come in, the houseboy used the event to say a good word for George.

Greenawalt got up, donned a robe, and went downstairs. He took some money from the paying customer and told the houseboy to show him to a room where some other men had left a vacant bed. Atzerodt sat up and asked if he could have his old room, number 51. It was occupied, Mr. Greenawalt said, but he was welcome to accompany the stranger and find a bunk in the same room.

The stranger registered as Sam Thomas. He gave five dollars and received change. Atzerodt tried to follow him up the stairs, but Mr. Greenawalt stopped him. "Atzerodt," he said, "you have not registered."

"Do you want my name?"

"Certainly."

The conspirator signed his name and went to the room. He had finally found a place to sleep.

3 a.m.

Now the nation slept. Tenth Street was deserted. Washington City was quiet. So were Ashtabula and Asbury Park. The few who were acquainted with the tragedy slept as soundly as the many who had yet to hear of it. Night trains roared through the countryside with wide-awake engineers, and milk wagons clanked to stores east and west and north and south, and policemen in fawn helmets yawned at street corners, but still the nation slept.

Even young Anna Surratt and Honora Fitzpatrick had long since exhausted the giggles in remembering the fierce detective who had pulled back the bedclothes a little. They too slept. Atzerodt slept. Major Rathbone slept. So did little Tad Lincoln, who had been in bed at the White House since 8 P.M.

This hour was the quiet one.

In Petersen House, Stanton decided that a second news bulletin should go at once to General Dix in New York. He still bustled at his work. Corporal Tanner yawned and pretended to be thinking, with eyes closed, of his notes. Justice Cartter sat, with crossed legs, staring out a back window which, by daylight, gave little view, and at night none. Stanton wanted to send this notice to Dix so that he could rectify an early mistake and name the assassin.

Washington City
No. 458 Tenth Street, April 15, 1865
3 a.m.
Major-General Dix,

(Care Horner, New York)

The President still breathes, but is quite insensible, as he has been ever since he was shot. He evidently did not see the person who shot him, but was looking on the stage as he was approached behind.

Mr. Seward has rallied, and it is hoped he may live. Frederick Seward's condition is very critical. The attendant who was present was stabbed through the lungs, and is not expected to live. The wounds of Major Seward are not serious. Investigation strongly indicates J. Wilkes Booth as the assassin of the President. Whether it was the same or a different person that attempted to murder Mr. Seward remains in doubt. Chief Justice Cartter is engaged in taking the evidence. Every exertion has been made to prevent the escape of the murderer. His horse has been found on the road, near Washington.

> *Edwin M. Stanton*
> *Secretary of War*

He sent this to Bates, at the War Department Telegraph Office at once. Little time was wasted on any of Stanton's messages. Troopers at the curb in front of Petersen House were given the dispatches, and rode at top speed down to E Street, across E to the south White House grounds, and up Seventeenth to the department. Here, young soldiers waited at the curb to take messages upstairs to Bates and, at the same time, to give messages to the troopers for Stanton.

It was an efficient system. The message above was sent over the wires at 3:20 A.M. By 4 A.M. it had been read to the New York press. The dispatch itself is significant only because it shows that Stanton was beginning to change his mind. He had begun with the notion that Washington was seething with assassins and arsonists; that a reign of terror had overtaken the city and death was to overtake many people before dawn.

Now, almost five hours later, he had a suspicion that the Federal Government was fighting one man. "Every exertion has been made to prevent the escape of the *murderer. His* horse has been found. . . ." If the new thesis was correct, then Stanton, with all the majesty and power of the United States Government behind him, was a damned fool. He had been outwitted, was being outwitted, and might continue to be outwitted by a lone actor. Because of this, and for no other reason, the Secretary of War would, in the days ahead, insist that this was all part of a huge conspiracy, inspired and approved by the defunct Confederate States Government. He could not admit, even to himself, that he was not battling Davis and Benjamin and Seddon and Stephens. It was big, or Stanton was ridiculous.

Speed wanted to leave the premises for a while, and he brought to Stanton the letter which would notify Johnson that the President had died. Stanton placed it on Corporal Tanner's table, and the young man read it:

Sir:

Abraham Lincoln, President of the United States, was shot by an assassin last evening at Ford's Theatre, in this city and died at the hour of——.

About the same time at which the President was shot an assassin entered the sick chamber of the Hon. William H. Seward, Secretary of State, and stabbed him in several places—in the throat, neck, and face—severely if not mortally wounding him. Other members of the Secretary's family were dangerously wounded by the assassin while making his escape.

By the death of President Lincoln the office of President has devolved under the constitution upon you. The emergency of the government demands that you should immediately qualify according to the requirements of the constitution, and enter upon the duties of President of the United States. If you will please make known your pleasure such arrangements as you deem proper will be made.

Your obedient servants,

Hugh McCulloch, Secretary of the Treasury
Edwin M. Stanton, Secretary of War
Gideon Welles, Secretary of the Navy
W. Dennison, Postmaster-General
J. P. Usher, Secretary of the Interior
James Speed, Attorney General

To Hon. Andrew Johnson,
Vice President of the United States.

At police headquarters, Major Richards wrote an order of small importance to any but drinkers:

Washington City, April 15, 1865
Three o'clock a.m.

In view of the melancholy events of last evening, I am directed to cause all places where liquor is sold to be closed this entire day and night. The sergeants of the several precincts are instructed that this order is enforced.

A. C. Richards,
Superintendent.

General Augur received a report of the raid on Surratt House before he could organize his forces, and the news about the Surratt family was forwarded to Stanton with the news about the stableman who serviced Booth's horse, the letter from Sam Arnold to Booth, and other late data including Atzerodt's peculiar behavior at Kirkwood House. For the first time, a tavern in southern Maryland, at Surrattsville, came into focus.

Stanton and Cartter went over each report with care, and the more the two men listened and read, the more it became apparent that they were battling two or three men—or at most, a half dozen—all of whom had an affinity for southern Maryland or for Baltimore. The best news of all was the letter from Sam Arnold to Booth, because that established, beyond argument, that there was a plot; it established that such a plot had existed for weeks; it established that Arnold thought that Booth should not move until he first heard from "R——d."

At once, the Secretary of War began to work on a new dispatch for General Dix.

4 a.m.

The steeple at Beantown was black against the night sky when John Wilkes Booth slowed his horse and Herold pulled the mare up. They were close to the home of Dr. Samuel Mudd, and whatever their plan was to be it had to be agreed upon now.

There is no record of the conversation between these two, but certain reasonable assumptions can be made from what happened. Booth did not regard Mudd as a friend. He put on the disguise of an old man.

Mudd was forty, tall, thin, had a bald forehead, blue eyes and brick-colored hair and whiskers. He was intelligent and independent. Until the Emancipation Proclamation, he had owned eleven slaves. Once, when a slave refused an order, Mudd drew a pistol and shot the man in the leg.

He owned a five-hundred-acre farm, and worked it. As was the case with his father and his brother, he liked property and he wanted more. He was a churchgoer and so was Mrs. Mudd. The doctor was influential in the neighborhood of Bryantown, and was a conservative Southerner in his politics.

Booth and Herold walked the sweat off their horses and talked about Mudd. The actor wanted to have the leg treated, and be gone. If it had to be bound, or splinted, all right. But Booth did not trust Mudd and, even though Booth was miles ahead of the news he had created, he knew that, if the doctor recognized him, sooner or later the Mudds would learn that Booth had killed Lincoln, and in that event the doctor had the type of character which would impel him to go to the authorities with the news. If Federal patrols were to come this way, looking for Booth, they would be here shortly after daylight—7 or 8 A.M. With luck, they might not reach this neighborhood until 10. But, once here, the story of the assassination and the search for Booth would be common gossip within an hour.

Booth, right now, was ten miles off his escape route. He was eighteen miles southeast of Surrattsville, when he should have been eighteen miles dead south. And, to get back to La Plata and down to Port Tobacco would now require the use of cross-country farm roads because there were no main roads. So, if Mudd could fix the leg so that riding would be bearable, they might get out of his house at 5 A.M. There would be some daylight then because sunrise would be 5:20.

The riding would be slow, but, if they made Port Tobacco by 7:30, they might still be ahead of the Federals, and if Atzerodt was waiting, as he should be, they would be moving out into Pope's Creek by 8 A.M. If Atzerodt wasn't waiting, they would have to hire a boat to take them across to Mathias Point. Of one thing Booth was certain: when they reached Virginia territory, and the great heroic news was known, every loyal Southerner would give them shelter and do them honor. John Wilkes Booth never doubted this, nor could he afford to, because once the civilian prop was removed from his future the actor was dancing on air. Herold, nodding to the superior wisdom of his worldly friend, believed with him. It did not occur to either of them that any Southerner could or would greet them with contempt.

They swung off the road and up before Dr. Mudd's house. They agreed that Davey Herold would do most of the talking. The boy dismounted, and gravel crunched underfoot. Somewhere, a hound dog bayed and, in other places, dogs took up the cry. Herold knocked. Dr. Mudd, in bed, heard it and resolved that whoever it was could knock twice more if the matter was important. Herold obliged. The doctor came downstairs in nightshirt, holding a candle, and, from behind the locked front door, inquired who was knocking.

"Two strangers riding to Washington."

Mudd opened the door and, in the pale light, saw a young man. His horse was tied to a tree out front and the young man was holding the reins of another horse on which a silent man sat. The young man said that he and his friend were riding to Washington and his friend had taken a bad fall. His leg was hurt.

The doctor handed the candle to Herold, and went out and helped the other man off the horse. When he learned that the left leg was injured, he got on that side of the man and got under his arm and helped him up the stone step into the parlor. Herold followed with the candle.

"You sit there," said Mudd, helping the man to a sofa, "and I'll get more light."

Mrs. Mudd was at the head of the stairs and she asked what the trouble was. "A man hurt his leg," said the doctor. "It may be broken." Mudd lit two lamps and then crouched in front of the horseman. He did not try to pull the boot off. He just pressed both sides of the foot and ankle until he felt a mass and saw the patient jump.

"I don't think you will get to Washington tonight," he said. He looked at the old man wincing in pain. Then he looked at the boy watching. "I would suggest," the doctor said, "that you come upstairs with me and let me have a look at that leg."

Only the eyes and cheekbones of the injured man were showing. He nodded slowly, and Herold picked up one of the kerosene lamps and Mudd picked up the other. Between them, they assisted the silent man up the stairs. Mudd sensed that this man's pain was acute, but that he was trying not to make an outcry.

In the guest room were two beds. Mudd helped the patient to fall on the near one. The doctor stood near the bed, looking now at the patient's face, and the man kept the muffler up on his chin, although the house was warm. The doctor asked a few questions about the injury, and the patient groaned and closed his eyes. The young fellow stood in the doorway.

Dr. Mudd stooped and tried to pull the boot off. It wouldn't come off. The silent man raised his head off the pillow.

"Please make haste," he said. "I want to get home and have this attended to by my regular physician."

The doctor noticed that, when the patient moaned, his left hand went to the small of his back. Mudd excused himself and went downstairs and got some heavy pasteboard and some paste. He wet the insides of the pasteboards and glued them together until he had several very firm splints.

When he got back upstairs, he took surgical scissors and made a vertical incision in the boot directly over the instep, and cut straight up. When he reached the top, he peeled the leather back and, tugging gently, removed the boot, then the sock. A lump of purpled flesh showed about two inches above the foot. After probing and manipulating, Doctor Mudd found a simple fracture of the tibia, and no fractures of adjoining bones.

At 4:45 A.M. Mudd had finished his examination and had applied the splint. The patient then complained of a pain in his back and said it caused him to have trouble with his breathing. He was sure, the silent one said, that he could not be moved right now. This seemed to startle the young man in the doorway, but he said nothing.

"You can stay here," the doctor said. Mudd had good powers of observation. His mental notes were: Man five feet ten inches high, pretty well made. I suppose he would weigh 150 to 160 pounds. His hair was black and worn long and seemed to curl. He had a pretty full forehead and his skin was fair. To me he seemed to be accustomed to an indoor, rather than an outdoor life.

Mudd went downstairs and awakened his colored man, Frank Washington, and asked him to take both horses to the stable and to make sure that they had hay and water. The doctor decided that, as long as he was fully awake, he would have breakfast and he invited the young man to join him. The young man, he found, was talkative. He was short and dark, to the doctor's eyes, and appeared never to have had a reason to shave.

He told the doctor that his name was Henston and that the injured man was Mr. Tyser. He prattled on, telling the doctor that he knew him, although they had not met before, and that he was well acquainted in this part of Maryland. Mudd found him to be guileless and superficial.

After breakfast, the doctor was about to go out into the fields when the young man asked if he could borrow a razor. The doctor, normally an apprehensive and suspicious man, had been at ease until now. He asked what the blade was to be used for, since he had noticed that the boy had no beard. Herold said that his friend upstairs would like to shave.

"It will make him feel better."

The doctor gave him a razor. Mudd's suspicions were aroused because upstairs when the shawl had slipped a little bit, he had seen part of a full graying beard and a coal-black mustache. It did not seem reasonable that a man in pain, in a stranger's house, should suddenly decide to remove a well-nourished beard or mustache. Dr. Mudd went out into the fields to direct the day's work.

Stanton finished his third press bulletin to General Dix, and because it supported his "huge conspiracy" feelings, he was pleased with it. It was sent by Bates at 4:44 A.M.:

Major General Dix:

The President continues insensible and is sinking.
Secretary Seward remains without change. Frederick

Seward's skull is fractured in two places, besides a
severe cut upon the head. The attendant is still alive
but hopeless. Major Seward's wounds are not dangerous.

It is now ascertained with reasonable certainty
that two assassins were engaged in the horrible crime,
Wilkes Booth being the one that shot the President,
the other a companion of his whose name is not
known, but whose description is so clear that he
can hardly escape. It appears from a letter found in
Booth's trunk that the murder was planned before
the 4th of March, but fell through then because the
accomplice backed out until "Richmond could be
heard from."

Booth and his accomplice were at the livery stable
at 6 this evening, and left there with their horses
about 10 o'clock, or shortly before that hour. It would
seem that they had for several days been seeking
their chance, but for some unknown reason it was not
carried into effect until last night. One of them has
evidently made his way to Baltimore, the other has
not yet been traced.

> *Edwin M. Stanton*
> *Secretary of War*

This was, in a manner of speaking, a defensive press release. When Stanton wrote, "It appears from a letter found
in Booth's trunk that the murder was planned before the 4th
of March," he lied. There is nothing in Sam Arnold's letter
that bears on inauguration day. The letter itself was mailed
on March 27. The only possible way that the Secretary of War
could have been reminded of a period prior to March 4 was if
Augur had told him *now* that Booth and Surratt and Atzerodt
had been named in a presidential kidnap plot prior to March

4 (by Wiechman and Captain Gleason) and that the department had done nothing about it.

Mr. Stanton had Arnold's letter before him when he wrote the dispatch to Dix, and he quoted Sam as writing to Booth that the accomplice backed out until "Richmond could be heard from." What Arnold wrote was "I would prefer your first query: 'Go and see how it will be taken in R——d . . .'" The misquotation must have been deliberate because there is an enormous difference between going to ascertain how Richmond will react to a scheme, as opposed to holding a plot in abeyance until "Richmond can be heard from."

He was wrong in other, lesser matters, but these were errors of judgment. Stanton now believed that Booth and Arnold were the culprits and, because Arnold's letter was dated from Hookstown, Balto. Co., he alerted the Baltimore Department to locate Samuel Arnold and arrest him in the Seward assassination. Mr. Arnold was working as a clerk at one of the War Department's bastions: Fortress Monroe, in Virginia.

At the same time, Stanton was convinced that Booth had escaped from Washington City, and, in the light of the reports he had on activities at the Surratt boardinghouse, he called for maps and called a conference of ranking officers.

It was late and almost ludicrous to be examining the bars of the municipal cage now, but it was done. Military reports from the north, from the west and from the south showed that no one resembling Booth had been seen on any of these roads. All of the city exits were examined and the only one uncovered, unwatched, unpatrolled, was the peninsula called southern Maryland.

One of the officers reminded the Secretary of War that the blind horse had been found in East Capitol, almost on the route to the Navy Yard Bridge and southern Maryland. If, the military minds reasoned, Booth went that way, then he would be bottled up in the area unless he could get back to Virginia.

His best chance to get back on Old Dominion soil would be in the region of Piscataway, Maryland, or below Indian Head. Stanton asked for a picture of Booth and someone got one from the files at Ford's Theatre. It was a picture of Edwin Booth.

Another troop of cavalry was ordered out and was told to patrol the area of Piscataway, Maryland, and if clues were turned up to signal the War Department by telegraph. It was commanded by Lieutenant David D. Dana, the younger brother of the Assistant Secretary of War. He was requested to cover the road north out of Piscataway, south to Accokeek, and northeast to Surrattsville.* Piscataway was a good junction of roads for such a search. If Booth was not in this area, and had not been seen, then he was probably headed for Annapolis or Upper Marlboro. Either that or—Stanton placed little credence in this—he was sleeping somewhere right here in Washington City.

---

* On modern maps, it is often listed as Clinton.

5 a.m.

A dark stain spread around the President's head and the Surgeon General announced that Lincoln had sustained a fresh hemorrhage. The doctors lifted his head and a new pillow and case were placed on the bed. The hair around the wound was cleaned and cotton batting was pressed against it. After that, the President's breathing appeared to be more regular.

Gray light began to swell against the bedroom windows. In the room, the gaslight seemed to pale. Dr. Barnes sat at the head of the bed, trying to take a pulse from the carotid artery. Halfway to the feet of the patient, Dr. Leale sat, still holding Lincoln's hand, now and then checking the pulse in the flaccid wrist, sometimes getting a count, sometimes getting nothing. On the wall side of the bed, Dr. Stone sat, as helpless as the others.

The faces in the room had changed during the long night. Two, besides the doctors, remained constant. One was Robert Lincoln, still standing behind the head of the bed, looking down. The other was Secretary of the Navy Welles, fat and solemn, sitting with a hand on one knee, staring at this man who was, in effect, almost the last soldier to die in the war.

Death was roosting in the room now. Everyone present knew it. In the early hours, the President looked relaxed and, in spite of the medical prognosis, one would expect him to awaken any moment. Now, with the bullet lodged directly behind it, the right eye was swollen and purpled. The lips were cyanotic. The heart throbbed, skittered, and seemed to stop. The legs were as cold as the marble tabletops. Breathing stopped for long periods, and after a few seconds one or two of the doctors would pull out watches to note the exact time of death. Suddenly, the lungs would burst with air, the heart would dance with life, and the President would groan through half-opened lips, as though, in a dream, he was walking down the White House stairs asking: "Who is dead?"

Mrs. Lincoln sat quietly. Laura Keene and Clara Harris were too spent for conversation. In the silence, they sat looking at the wall, or watching the inexorable growth of light in the room. As in a distance, they could hear the deep voices of the doctors in the bedroom and, now and then, the thin pitch of Mr. Stanton asking for something. In sound, the thud of boots and the clank of spurs never seemed to stop. For days, it seemed, soldiers had been walking through the hall outside this room. For the rest of her life, Mrs. Lincoln would dread the sight and sound of them.

An officer of General Augur's staff awakened the general agent of the Baltimore & Ohio Railroad at his hotel. This was Mr. George S. Koontz, and he did not know the news. When he heard it, he dressed quickly, asking again and again to make sure that this was not a mistake or, worse, a joke.

The captain accompanied Mr. Koontz to the terminal and said that the government wanted the B & O to stop all trains leaving Washington. All road exits from the city had been sealed, and Stanton wanted to prevent the assassins from leaving town by train.

At the depot, they found the waiting room and the train platforms swarming with soldiers and detectives. The first northbound train scheduled was the 6:15. Koontz issued orders that no train was to leave the station. When the cars had been made up, and backed into the siding, passengers climbed aboard and then detectives went aboard and studied every person in every car. The soldiers examined all luggage and all mail bags. They questioned the engineer, the fireman, the conductor and the brakemen. It was then decided, although the scheduled time had not arrived, that this train could depart at once.

The bewildered passengers stared out of the windows as the train eased out of the depot, picked up speed, and got as far as Relay House, a short distance on the road to Baltimore, where it was stopped by General Tyler. He explained to the train crew that Relay House was in his domain, and that if the army in Washington permitted trains to leave the city, that was their business. His business was to stop them at Relay House, and there the trains would remain.

The general, who got his star by obeying orders implicitly, also stopped all southbound trains out of Baltimore so that, in a short while, he had several trains standing on the tracks. Some passengers begged him to permit two sick children to continue their journey home, but the general said no.

Thus, if any of the conspirators had elected to take the morning train out of Washington, he would have been stopped at Relay House—provided, of course, that he could get through the tight military net at Washington depot—and later, when traffic was cleared through to Baltimore, such a conspirator would not have been able to make any train connections to New York or to Canada, as some were to charge that John Surratt did.

It was another gray and misty day and Robert Nelson, Negro, was walking across Lafayette Square on his way to work. He was crossing the street in front of Mr. Seward's home when he saw a knife. He picked it up and turned it over in his hand. A soldier, now patrolling in front of The Old Clubhouse, watched him and came out in the street and asked what he had picked up. Nelson showed him the knife. The soldier took it.

Lewis Paine had dropped it.

Twenty minutes later, a half mile to the east, William Clendin was walking down F Street toward Eighth when he saw a Negro woman run out of a doorway, step into the gutter, and pick something up. As he approached, he asked her what it was. In silence, she handed him a knife and a sheath.

A woman leaned out of an upstairs window and told Clendin that she had raised the shade and had noticed something in the road and had sent her maid down for it. Clendin held it aloft and told her that it was a knife and sheath. The lady said she would not permit it in the house and slammed the window.

Clendin turned it over to the police. Atzerodt had thrown it away.

6 a.m.

C Street glistened with mist. The Pennsylvania House looked a little bit more dismal than usual as George Atzerodt came out, looked up and down, then crossed to the other side and started up Sixth Street. He was sleepy, and dirty, and penniless. He had a hangover. He was sick, soul and bone. The morning air was chill. He dug his hands into his trouser pockets and walked up toward the Mall.

"Mr. Atzerodt. What brings you out so early?"

The carriage maker jumped. He looked up. A colored boy from the hotel was coming back after seeing a lady guest off on the morning train.

Atzerodt's grin was forlorn. "Well," he said, "I have got business."

Mr. Atzerodt had business all right. He wanted to hide. In bed he had thought of many places, and now he had made up his mind to hide in the little town where he had first started in America. He knew that this was not a good place in which to hide, but he reasoned that, no matter where he hid, no matter how far away, the news would reach that place and they would come and get him. The world dealt harshly with cowards. A judge would not believe that he, George Atzerodt, could not kill anyone. So he had to hide. And, not having any money with which to ride, this conspirator was going to walk.

He would walk westward, through Washington City, through Georgetown, until he got to his little town. There, the people liked George Atzerodt. They were not like the people of Port Tobacco. They knew him as a harmless buffoon, a beaming, perspiring drunkard. He would hide there, listening and laughing and maybe drinking until some men came and asked him if he was George Atzerodt.

When he got to the Mall, he turned on Constitution Avenue and he walked and walked and walked, the furtive piggish eyes dancing, the dampness on his round hat. This was a stupid conspirator. He was doing what no other conspirator would—walk through the enemy lines the day after the high crime. It wasn't brazenness. Nor courage. The man had no other place to go.

In Surratt House Louis Wiechman had breakfast with Mr. Holahan. Wiechman was talkative and Wiechman was righteous. The police would have raided the place a long time ago if they had listened to him. He had suspected what was going on and if Booth plotted this dangerous thing, and John was foolish enough to get into it, then John deserved whatever he got out of it. Thank God that he, Louis Wiechman, had not become part of it; had, in fact, gone on record as reporting his suspicions months ago.

Now, today, he was going to do his duty as any self-respecting citizen should. He was going down to police headquarters right after breakfast and he was going to offer his services to the police. He would help them to track down his dearest friend, John Surratt, no matter whether the trail led across the Eastern Branch to Surrattsville or up north to Canada.

Holahan pushed his plate away. He had little to say. He might have reminded Louis that he had heard more secessionist talk from him than from the others. He didn't.

Mr. Holahan stood and said that he had some things to do. Wiechman finished the breakfast Mrs. Surratt had prepared for him and then walked off to the police station. He had an excellent memory and he could quote old dialogue as though it had been uttered yesterday. He would talk and talk and talk until the police tired of listening.

He was a hanging witness.

A few streets away, James Ferguson was finishing breakfast when Mr. Gifford, chief carpenter at Ford's Theatre, walked in looking irritated.

"You made a hell of a statement last night," he said. "How could you see the flash of the pistol when the ball was shot through the door?"

Ferguson was puzzled. The stage carpenter said that the authorities had discovered that Booth had fired the fatal shot through the door of Box 7, and he had seen the hole to prove it.

"Mr. Gifford," said Ferguson fervently, "that pistol never exploded in any place but the box. I saw the flash."

"Oh hell!" said Gifford walking out. "The ball was shot through the door. How could you see it?"

Old Gideon Welles had sat with his President as long as he could. He needed a stretching of aging limbs, a breath of air. He got his coat and, when he put it on, fluffed his white whiskers outside the lapels, jammed a broad-brimmed felt hat over his long brown curls, and walked outside.

The chatting sentries on the walk snapped to attention. Mr. Welles walked slowly around the block, noting that small groups of people huddled against the buildings in the drizzle. They were waiting for news. Final news.

They recognized the old Secretary of the Navy and some looked at him expectantly, but he said nothing. Sometimes, a person would ask timidly: "Is there no hope?"

Once or twice, he said "No," and kept walking. Once he said, "The President can live but a short time." He was affected by the colored people, who stared at him, unable to ask the question except with their eyes. He noted that there were more of them standing this death watch than white. On some faces, he saw the varnished furrow of old tears. To one group, without being asked, he was so moved that he said it would be better now if the President did not live.

Mr. Welles finished his tour and walked up the small steps at Petersen House and went back into the bedroom. Outside, two sentries made a game of trying to remember the names of all the important people who had come to this house this night. They remembered thirty. There were 46: Hugh McCulloch, Edwin M. Stanton, Gideon Welles, John P. Usher, William E. Dennison, James Speed, Andrew Johnson, William T. Otto, Robert King Stone, Joseph K. Barnes, Thomas T. Eckert, John B. S. Todd, Schuyler Colfax, Robert T. Lincoln, Charles Sumner, Maunsell B. Field, Leonard J. Farwell, Isaac N. Arnold, C. H. Liebermann, Charles H. Crane, John F. Farnsworth, John Hay, Gilman Marston, David K. Cartter, J. C. Hall, Charles S. Taft, Christopher C. Augur, Charles A. Leale, Henry R. Rathbone, Almon F. Rockwell, Louis H. Pelouze, E. L. Dixon, Thomas M. Vincent, Clara H. Harris, Constance Kinney, Richard J. Oglesby, Edward H. Rollins, Montgomery C. Meigs, Mary C. Kinney, Isham N. Haymes, Benjamin B. French, Phineas D. Gurley, George V. Rutherford, Lyman B. Todd, Henry Halleck, Laura Keene.

Now a new sound could be heard. Newsboys, carrying the morning papers, were shouting the bulletins up and down Tenth Street and the soldiers out front were buying the papers. The story, running down the left side of page 1, looked like this:

IMPORTANT

ASSASSINATION

President Lincoln

The President Shot at The Theatre Last Evening

Secretary Seward
Daggered in his Bed
But
Not Mortally Wounded

Clarence and Frederick Seward
Are Badly Hurt

Escape of the Assassins

Intense Excitement In
Washington

At one of the police precincts, the man nobody missed showed up. John F. Parker, unseen since ten o'clock last night, walked into the station with his contribution to justice. He had a prostitute by the arm and he told the sergeant of her crimes. She was Lizzie Williams. The sergeant looked her over. She was scared and drunk. He shrugged and refused to book her. She was ordered to get out of town.

Parker did not offer to tell the sergeant where he had been all night, and the sergeant did not ask. The policeman did not ask the condition of the President, nor did he offer to file a report about the assassination. The sergeant advised him to go home and to get some sleep. Parker left. He remained a policeman in good standing for three more years. He was not tried and no charges were filed against him.

Corporal Tanner had finished his work for Secretary Stanton and he stood, picked up his notebooks, crushed his hat under his arm and, in leaving, passed through the bedroom where the President was. His uniform was rumpled and baggy and he tiptoed to the bed, standing between Generals Halleck and Meigs. He looked at the face that would be no more, and he studied the two doctors who sat on the bed.

Nobody looked at Corporal Tanner; no one spoke to him. He watched Stanton, at the foot of the bed, pressing a fist into a palm. Robert Lincoln, dry-eyed, stood behind the bed. The corporal put his uniform cap on, saluted, and walked out.

7 a.m.

John Cass, merchant, in Elmira, New York, started his day thinking—after a look out of the bedroom window—that this was going to be one more cloudy Saturday. It was going to be a bad business day too, because first of all money was still tight, and secondly, no one was going to buy a suit of spring clothes in Cass's Clothing Store so long as the weather remained cool.

Then the news reached Mr. John Cass. It came at the breakfast table and it hurt. Everybody, it seemed, except Cass knew that the President of the United States had been shot last night, the Seward family had been murdered in its beds, murderers were running loose in Washington City. Cass looked at his eggs, the homemade bread and the jelly, and he said that he must go to the store at once; it must be closed for the day.

He did not weep, as so many did this morning. He was too stunned for realization. He got his coat and he went out and walked down to the corner of Water and Baldwin Streets trying to convince himself that Lincoln had been shot and was dying, was, perhaps, dead by now. First he stopped across the street at the telegraph office and asked if there was any further news about the President. The telegrapher shook his head. The last news on the key, hours ago, had said that the President could not live.

John Cass walked across the street, unlocked his store, and got a big piece of cardboard. He was going to handletter a sign explaining that the store would remain closed out of respect to Mr. Lincoln. He looked up, trying to marshal the wording, and he saw a young man coming across the street. His attention was fixed on the fellow because he was wearing a type of coat seldom seen in Elmira. It was called a Canadian coat. The young man walked in and asked for a white shirt. He was pleasant, and had a nice smile. He was fairly tall, had a domed forehead, and wore a faint, wispy goatee. He asked for a very special style of white shirt. Mr. Cass told him that he was sorry; he did not keep that particular brand, but he had others just as good. In a moment, the sign had been temporarily forgotten, and the urge toward business had stepped forward.

Cass brought out a number of fine white shirts. The customer said no, he'd prefer to wait until he could buy that particular brand. "Well," said Cass, "you won't find any in this town." He was putting the shirts back in the bins and he said: "We have received some bad news."

"What's that?" the customer said.

"The death of Abraham Lincoln."

The young man made a discourteous remark, which Mr. Cass never again repeated. However, he stared at the fellow and would remember him forever. This was John Surratt. His interest in Elmira, New York, on that morning was that 5,025 Confederate prisoners were held there.[*]

Heavy pelting rain began to fall in and around Washington City, and at Piscataway Lieutenant David D. Dana made his first report by telegraph on the hunt for Booth:

Captain R. Chandler,
Assistant Adjutant-General
    Sir: I have the honor to report that I arrived in this place at 7 A.M. and at once sent a man to Chapel

---

[*] Surratt was later identified by Cass and several others in and around Elmira as having been in the town on the morning of April 15, 1865. Had he been in Washington the night before, there was no possible way he could have reached Elmira in the morning.

Hill to notify the cavalry at that point of the murder of the President, with descriptions of the parties who committed the deed. With the arrangements which have been made it is impossible for them to get across the river in this direction.

Very respectfully, your obedient servant,

David D. Dana,
First Lieutenant and Provost-Marshal, Third Brigade.

I have reliable information that the person who murdered Secretary Seward is Boyce or Boyd, the man who killed Captain Wilkins in Maryland. I think it without doubt true.

D. D. Dana.

He was wrong, but he was young and he was trying. As he wrote, John Wilkes Booth slept soundly, the painful leg forgotten, in Dr. Mudd's house twelve miles to the southeast. Sixteen miles straight north, in a patch of Washington woods, Lewis Paine crouched, wondering how long he could hold out and how one proceeds to get out of the city, and where "Cap" had gone. On the upper reaches of Wisconsin Avenue, in Georgetown, George Atzerodt walked more slowly. He sold his gun for ten dollars, and now, if he could find a place in this farmland where drinks were served, he could buy one. Downstairs in Doctor Mudd's house, David Herold learned that even assassins cannot stay awake forever. Sleep stilled him in a chair.

President Lincoln's right eye was black. He began to moan—the long, frightening moans heard so often by the guard Crook in the White House corridor—and some around the bed in Petersen House felt that he had endured pain so long that he could no longer bear it in silence. His breathing became shallow and swift. The lips blew outward and were sucked in. His black hair was in disorder and there was a slight ruddiness on the cheekbones.

The big veinous hands were composed on the white sheet, and the feet stuck out into the aisle.

In the back parlor, his old friend of the circuit-riding days, Interior Secretary John Usher, snored on a couch. Justice Cartter sat in silence, looking out a window at the rain.

Surgeon General Barnes looked at the other doctors, felt the cold skin of the patient, and asked an officer to bring Mrs. Lincoln to the bedroom. Robert Lincoln heard the words and buried his face in his hands. Stanton came in and stood at the foot of the bed, his hat in his hand.

They brought Mrs. Lincoln in, tottering, and she looked at her husband and, hearing her son sob, looked at him. Then, without a word, she was led out of the room. Secretary Welles came into the room, sat, and then stood again. The President was in his death struggle and, at intervals, he tried to breathe but all that happened was that he pulled the cheeks inward but the lips remained closed.

Surgeon General Barnes studied his watch. Dr. Phineas Gurley, the President's pastor, came in from the front parlor and looked at the thin red rug. Dr. Leale saw the chest heave upward, hold the position, and then relax. The time was twenty-two minutes and ten seconds past 7 A.M.

Dr. Barnes stood, waiting for the next breath. It did not come. He peeled an eyelid back, looked closely, and pulled down the sheet and listened with his ear against the plaster-covered chest. He remained in this attitude for some time, then he straightened up, reached into his vest pocket, and withdrew two silver coins. He placed them on the President's eyes.

Secretary Stanton broke the silence. "Now," he said, "he belongs to the ages." He clapped his hat on his head. Someone whispered that Dr. Gurley would say a prayer. The hat came off. As the minister began the slow, soft prayer, Robert left the room to tell his mother.

Leale performed the last loving service. He composed the arms of the man. Dr. Gurley knelt on the floor and said: "Let us pray." The men of the Cabinet were standing with bowed heads as Robert led his mother into the room. She tore loose and threw herself on her husband and cried: "Oh my God! I have given my husband to die!"

She was lifted from the bed and taken from the room again. This time she was partly carried. Leale smoothed the skin of the face back toward the ears. He removed the silver coins, smoothed the eyelids gently closed, and replaced the coins. He pulled the white sheet over the head.

He said nothing. Dr. Gurley was still praying. Dr. Leale got his uniform coat and walked out, putting it on. He was in a haze of fatigue and knew only vaguely that he was walking in heavy rain. He put his hand to his head and noticed that he was hatless. Then he remembered: the hat was still in Ford's Theatre.

The doctor did not go back. He wanted to walk. He was walking—he knew not how long—when he heard the first deep toll of the bell. It came from nearby and the bass sound reverberated a long time before it died in resonance and a second toll sounded. In a few minutes, he heard another bell, a higher-pitched bell, and the sound clashed with the first one. Then he heard another, and another, and soon, as he walked, the rain dripping off his forehead, Dr. Leale was surprised to learn that Washington City had many bronze tongues.

The bells were tolling everywhere, it seemed, and people came out on the street, heedless of rain, to listen. No one asked what the bells meant or, if someone did, it was not recorded that he did. These people were trying to believe something which was hard to believe. Some looked sick. Some were grim. Some swore loud oaths to a wet sky. Some wept. Patrols called a halt to stand and listen. A barber on F Street bent to his task, looked at the hand holding the razor, and folded it up. "No more," he said to the customer. "I'm sorry, but no more now."

Telegraph keys began to chatter north, east and west and bells began to toll in Boston and Chicago and in Springfield, Massachusetts as well as in Springfield, Illinois.

Gentle men were trying to lead Mrs. Lincoln out of the house, and she kept trying to squirm from their grasp. She was shouting: "Oh, why did you not tell me he was dying!" As she was led down the steps of the stoop, she saw the brick façade of Ford's Theatre across the street. "Oh!" she screamed. "That dreadful house! That dreadful house!"

Inside, Stanton wrote:

Washington City April 15, 1865
Major General Dix,
New York:

Abraham Lincoln died this morning at 22 minutes after 7 o'clock.

Edwin M. Stanton
Secretary of War

Then he too left. And Welles left. And Speed and Usher and Cartter and all the rest. General Vincent remained with the body. He told an officer to go get a closed hearse and an honor guard of soldiers. They would walk the body back to the White House.

As if the bells were not enough, big-mouthed cannon all over the nation began to boom every thirty minutes all day and all night. The requiem was in bad taste. Strangest of all, millions of people who had not cared much one way or the other now discovered that they loved this man. Maybe they loved him only because he was a martyr. The tears were universal. In New York, a red-eyed man, sober, stood on a corner talking to no one in particular and he said: "If he could just come back for one moment, I know what he'd say; he'd say 'Forgive him—he knew not what he did.'"

In Coles County, Illinois, the news came and farmers hitched up their buckboards and drove en masse across cold-looking fields to the little place where Sarah Bush lived; she was Lincoln's stepmother. They stood on the doorsill and they told her the tidings and her old leathery face did not change when she said: "I knowed when he went away that he would never come back."

Across the street from the White House, on the far side of Pennsylvania Avenue, the plain people waited to say good-by. Mostly, they were Negroes and they formed a thick dark ribbon on the walk. The cold rain stitched their backs but they did not move. The men wept too, and one called out: "If death can come to him, what will happen to us?"

The rain beat hard against the White House portico and, inside, the Secretary of the Navy shook drops from his hat. He looked up and saw little Tad Lincoln, with strained dignity, coming down the stairs.

"Mr. Welles," he said, "who killed my father?"

And that night, history has it that another little boy, whose name is lost in anonymity, sat chattering in the cold on a buckboard beside his father. All day long he had heard men, his father included, repeat the story that shook them to tears. All day long he had been frightened by the slow tolling of the bells and the smashing roar of the big guns; all day long, without asking questions, he had watched women tack bolts of black around their front doors and now, as he looked up into the cold sky, his childish heart could not believe that the stars were out.

He just couldn't believe it.

Postscript

Perhaps, like me, you wonder what happened to some of the people who played a part in this day of April 14, 1865. I will tell you, in no set order, the little that I know:

Surgeon General Barnes lived long enough to minister to the assassinated President Garfield. Andrew Johnson lived ten years. William H. Seward recovered, and died in 1872 of natural causes. He was seventy-one. James Speed, Lincoln's old friend, resigned as Attorney General in a year.

Stanton was forced from his post by Johnson and begged to be appointed to the Supreme Court of the United States. The appointment arrived as he was on his deathbed in 1869. The Secretary of the Interior, John Usher, resigned in a month. Gideon Welles lived to be seventy-six years old. General Augur, who had been in Grant's class at West Point, retired from the army and lived to see the start of the Spanish-American War. Schuyler Colfax became Vice President of the United States and was later involved in the Credit Mobilier scandal. William H. Crook, the guard, lived a great number of years and wrote his memoirs.

Thomas Eckert, who could break pokers over his arm, became a general, retired, became head of a big commercial telegraph company, and lived until 1910. The owner of Ford's Theatre, John T. Ford, was thrown into prison, but was later released for lack of evidence. The government confiscated his theater, but he forced it to pay $100,000 for the house. Twenty-eight years later, the floors of Ford's collapsed, killing more than a score of government workers. Today, rebuilt, the theater is a national museum.

Ulysses S. Grant, in time, became President of the United States, had a poor term of office, became a tool of Wall Street operators, and wrote extensive memoirs to keep from dying penniless in 1885.

Bessie Hale, the Senator's daughter who loved John Wilkes Booth, later married William Eaton Chandler, who was not an actor. Clara Harris was killed by her husband, Major Henry Rathbone, who, in turn, lived out his days in an insane asylum. Marshal Ward Hill Lamon, who might have saved Lincoln, regretted all his days (and they covered the next twenty-eight years) that he was in Richmond the night the President was shot.

George Atzerodt was caught, tried and hanged. So were Lewis Paine and David Herold. Booth was cornered in a Virginia barn and shot. For years afterward there were stories that it wasn't Booth who was shot, but the stories were wrong. It *was* Booth and, years later, when the government removed his body from under a stone floor in a prison, and sent it home, the Booth family identified the remains as those of John Wilkes Booth and buried him in the family plot.

Mrs. Mary E. Surratt was tried, convicted and hanged for conspiracy. On a hot July day, a government employee held an umbrella over her head before the trap was sprung. On the morning of the hanging, her daughter Anna tried to see President Johnson to beg for mercy for her mother. Anna was kept from seeing the President by Preston King of New York and Senator James H. Lane of Kansas. Six months later, King tied a bag of shot around his neck and jumped off a Hoboken ferry; eight months after that, Senator Lane shot himself.

Dr. Samuel Mudd was tried for conspiracy and convicted. So were Sam Arnold, Mike O'Laughlin and Ned Spangler, the horse holder. All four were sentenced to Albany (New York) Penitentiary. Secretary Stanton, who felt that they had got off lightly, removed them to Fort Jefferson, Dry Tortugas

Prison, off Key West, Florida. There, in August, 1867, yellow fever broke out and, when the prison doctor died, Dr. Mudd volunteered his services. He saved the lives of soldiers and prisoners, but Mike O'Laughlin died. The officers of the post appealed for a pardon for Mudd and it was granted in February 1869. Arnold and Spangler were freed with him and, realizing that Ned Spangler was dying of tuberculosis, Dr. Mudd took him home to Bryantown with him, and cared for him until he died.

John Lloyd and Louis Wiechman became the government's star witnesses against Mrs. Surratt. Lloyd claimed he was threatened with death unless he testified against her. Wiechman claimed that Stanton promised him a job for his work as a witness, and for a time he worked in the Philadelphia customs house. He was later fired. When he died, he kept repeating that he was on his deathbed and he would still say that he told the truth at the trial of Mrs. Surratt.

John Surratt ran to Canada, thence to Europe, and was discovered two years later working as a Zouave forty miles from the Vatican. He was brought back, tried, and eventually released. He made money giving lectures on the assassination of Lincoln.

Mrs. Mary Todd Lincoln, perhaps the most pathetic of all the people who figured in this day, was certified as a "lunatic"* in Cook County, Illinois, ten years after the death of her husband. It was Robert's sad duty to sign the commitment papers. She was released a year later, and spent the last months of her life (1882) in a darkened room dressed in widow's weeds. In 1871, Tad died.

The last of the survivors, Robert Todd Lincoln, died at the age of eighty-three, in 1926.

---

* This word, in a time of psychiatric ignorance, was used to describe most emotional disturbances.

Bibliography

This is a list of the sources of information consulted before writing this book:

ANGLE, PAUL, *THE LINCOLN READER*

BROOKS, NOAH, *WASHINGTON IN LINCOLN'S TIME*

BROWNING, MARY E., *OUR NATION'S CAPITAL*

CLARKE, ASIA BOOTH, *THE UNLOCKED BOOK, A MEMOIR OF JOHN WILKES BOOTH*

DE WITT, DAVID MILLER, *THE ASSASSINATION OF ABRAHAM LINCOLN*

———, *THE JUDICIAL MURDER OF MARY E. SURRATT, DICTIONARY OF AMERICAN BIOGRAPHY*

EISENSCHIML, OTTO, *WHY WAS LINCOLN MURDERED?*

———, *IN THE SHADOW OF LINCOLN'S DEATH*

FERGUSON, W. J., *I SAW BOOTH SHOOT LINCOLN*

GRANT, U. S., *PERSONAL MEMOIRS*

HARNSBERGER, CAROLINE T., *THE LINCOLN TREASURY*

HERNDON, WILLIAM, *THE HIDDEN LINCOLN (HERNDON'S LETTERS)*

HURD, CHARLES, *WASHINGTON CAVALCADE*

———, *THE WHITE HOUSE BIOGRAPHY*

KIMMEL, STANLEY, *THE MAD BOOTHS OF MARYLAND*

LAUGHLIN, CLARA E., *THE DEATH OF LINCOLN*

LEECH, MARGARET, *REVEILLE IN WASHINGTON*

LEWIS, ETHEL, *THE WHITE HOUSE*

MEREDITH, ROY, *MR. LINCOLN'S CAMERAMAN*

MOORE, BEN PERLEY, *THE ATTEMPT TO OVERTHROW THE GOVERNMENT BY ASSASSINATION OF ITS PRINCIPAL OFFICERS* (3 VOL.)

MOORE, GUY W., *THE CASE OF MRS. SURRATT*

MORROW, HONORE, "LINCOLN'S LAST DAY," *COSMOPOLITAN MAGAZINE*, FEBRUARY 1930

MUDD, NETTIE, *LIFE OF DR. SAMUEL A. MUDD*

PITTMAN, BEN, *THE ASSASSINATION OF PRESIDENT LINCOLN AND THE TRIAL OF THE CONSPIRATORS*

PRATT, FLETCHER, *STANTON*

RUGGLES, ELEANOR, *PRINCE OF PLAYERS*

BIBLIOGRAPHY

SANDBURG, CARL, *MARY LINCOLN—WIFE AND WIDOW*

————, *THE PRAIRIE YEARS* (2 VOL.)

————, *THE WAR YEARS* (4 VOL.)

SEWARD, F. W., *REMINISCENCES*

STARR, JOHN W., *LINCOLN'S LAST DAY, TRIAL OF JOHN H. SURRATT IN THE CRIMINAL COURT FOR THE DISTRICT OF COLUMBIA (2 VOL.)*, **GOVERNMENT PRINTING OFFICE**

STERN, PHILIP VAN DOREN, *THE MAN WHO KILLED LINCOLN*

WILLIAMS, BEN AMES, *MR. SECRETARY*

WILSON, FRANCIS, *JOHN WILKES BOOTH*

WINSTON, ROBERT, *ANDREW JOHNSON*

EXHIBITS, FORD'S THEATRE AND PETERSEN HOUSE

GOVERNMENT TRACTS, WAR DEPARTMENT

NEWSPAPERS, NEW YORK PUBLIC LIBRARY

WASHINGTON *EVENING STAR*, CONGRESSIONAL LIBRARY

WASHINGTON NATIONAL *INTELLIGENCER*, CONGRESSIONAL LIBRARY

Index

ALSO BY JIM BISHOP

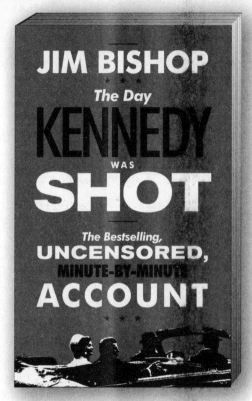

THE DAY KENNEDY WAS SHOT
The Bestselling Uncensored, Minute-by-Minute Account

Available in Paperback

Jim Bishop's trademark suspenseful, hour-by-hour storytelling style drives this account of an unforgettable day in American history. His retelling tracks all the major and minor characters of that day—JFK, Oswald, Ruby, Jackie, and more—illuminating a human drama that many readers believe they know well. As gripping as fiction but with a journalist's exacting detail, *The Day Kennedy Was Shot* captures the action, mystery, and drama that unfolded on November 22, 1963.